Fa

PITKIN COUNTY
library
inspire growth

PITKIN COUNTY LIBRARY
1 13 0003820879

D0450424

# Fabricated

## The New World of 3D Printing

Hod Lipson

Melba Kurman

John Wiley & Sons, Inc.

**Fabricated: The New World of 3D Printing**

Published by
John Wiley & Sons, Inc.
10475 Crosspoint Boulevard
Indianapolis, IN 46256
www.wiley.com



Published simultaneously in Canada

ISBN: 978-1-118-35063-8

ISBN: 978-1-118-41024-0 (ebk)

ISBN: 978-1-118-41694-5 (ebk)

ISBN: 978-1-118-63551-3 (ebk)

Manufactured in the United States of America

10 9 8 7 6 5 4 3 2



For general information on our other products and services please contact our Customer Care Department within the United States at (877) 762-2974, outside the United States at (317) 572-3993 or fax (317) 572-4002.

Wiley publishes in a variety of print and electronic formats and by print-on-demand. Some material included with standard print versions of this book may not be included in e-books or in print-on-demand. If this book refers to media such as a CD or DVD that is not included in the version you purchased, you may download this material at http://booksupport.wiley.com. For more information about Wiley products, visit www.wiley.com.

**Library of Congress Control Number:** 2012954401

*To Tiina and Jüri Kurman*

*To Rina and Stephen Lipson*

# About the Authors

**Hod Lipson** is a professor of engineering at Cornell University in Ithaca, New York. His work on automatic design and manufacture of robotic life forms, self-replicating robots, food printing, and bioprinting has received widespread media coverage including *The New York Times*, *The Wall Street Journal*, *Newsweek*, *Time*, and NPR. Lipson has co-authored hundreds of papers and speaks frequently at high-profile venues such as TED and the National Academies. Hod directs the Creative Machines Lab, which pioneers new ways to make machines that create, and machines that are creative. Visit `http://www.mae.cornell.edu/lipson`.

**Melba Kurman** is a technology writer, analyst, and popular blogger. She became interested in additive manufacturing and emerging design software two years ago when she noticed that the same manufacturing machine could print custom dental crowns, coffee tables, and heavy-duty, end-use metal machine parts. Her passion is explaining the value of complex technologies in everyday language. In the past she helped Cornell University, Microsoft, and other organizations bring new technologies to market. Melba is a graduate of Cornell University, the University of Illinois, and the U.S. Peace Corps.

# Credits

**Acquisitions Editor**
Mary James

**Senior Project Editor**
Kevin Kent

**Technical Editor**
Duane Storti

**Production Editor**
Kathleen Wisor

**Copy Editor**
San Dee Phillips

**Editorial Manager**
Mary Beth Wakefield

**Freelancer Editorial Manager**
Rosemarie Graham

**Associate Director of Marketing**
David Mayhew

**Marketing Manager**
Ashley Zurcher

**Business Manager**
Amy Knies

**Production Manager**
Tim Tate

**Vice President and Executive Group Publisher**
Richard Swadley

**Vice President and Executive Publisher**
Neil Edde

**Associate Publisher**
Jim Minatel

**Project Coordinator, Cover**
Katie Crocker

**Compositor**
Kate Kaminski,
Happenstance Type-O-Rama

**Proofreader**
Josh Chase, Word One, New York

**Indexer**
Johnna VanHoose Dinse

**Cover Designer**
Ryan Sneed

**Front Cover Product Design**
netfabb GmbH

**Front Cover Image**
© Paulo Kiefe, Creative Tools

**Back Cover Images**
© Objet Inc. and Olaf Diegel

# Acknowledgments

This book grew organically from illuminating and inspiring conversations with many people. We'd like to thank the following people for taking the time to speak with us and to review the book's content.

Duane Storti for his technical expertise and all-around useful insights; Robert Schouwenburg for his passion for all facets of 3D printing: technology, design, and business strategy; Timothy Weber, for the elucidating discussions, good laughs and "A-ha" moments; Eric Haines for thoughtful feedback and valuable insights; Mark Ganter for endless creativity in pushing the limits of the technology; Cathy Lewis, Rajeev Kulkarni, and Abe Reichtental, for hosting us at 3D Systems and sharing their vision; Brandon Bowman for discussions on food printing; Gonzalo Martinez for shedding light on the future of CAD; Adam Mayer for showing us around MakerBot; Bill Young and Ted Hall for sharing their first-hand experience in personal manufacturing; John T. Lee for showing us the myriad finer points of the commercial 3D printing process; Adrian Bowyer for hosting us at the University of Bath and his insights into intellectual property, the future of additive manufacturing, and for leading the RepRap revolution; Philip Delamore, Anthony Ruto, Ross Barber, and Hoon Chung for hosting us at the London College of Fashion; Jeroen van Ameijde for hosting us at the Architectural Association School of Architecture, London; Glen Bull, Jennie Chiu, William Kjellstrom, and Jake Cohen from the University of Virginia for teaching us about teaching; Jonathan Butcher and Yaser Shanjani for speaking to us about bioprinting; Chris Johnson for elucidating the future of CAD; Dave White and Ryan Cain for sharing their experiences with 3D printing in the classroom; Jenny Sabin for showing us her work of bio-inspired architecture; Jesse Roitenberg for enlightening us about Stratasys's involvement in education; Josh Harker and Eyal Gever for talking to us about their pioneering 3D printed art; Eduardo Napadensky, Ofer Shochet, and Daniel Dikovsky for inviting us on many visits to Objet and sharing with us the future of digital materials; Terry Wohlers for great research and analysis and sharing his data; Chuck Hull, Joe Beaman, and Carl Deckard for founding the field; Joris Peels for introducing us to Shapeways and Materialise; Michael Guslick for challenging our presumption on what should be printed; Noy and

Maor Schaal for daring to make some of the first printed food in history; Franz Nigl and Jeff Lipton for designing and creating elaborate multi-material cookies. And finally, to netfabb GmbH for designing the 3D printed hamburger on this book's cover, and to Paulo Kiefe of CreativeTools for freely sharing the image. The cover photo shows an art piece designed by netfabb using Selective Space Structures. The part takes advantage of the freedom in form and function offered by additive manufacturing. By mimicking the way nature uses structures the burger combines many different types of structures in one part to achieve different properties from a basic material. The piece was built using laser sintered PA by FIT Fruth Innovative Technologien GmbH in 2009.

We would also like to thank all of the designers, artists, and photographers who freely shared their images, illustrations, and pictures with us for the book's cover and interior.

Many thanks to the Fab@Home team in its entirety: its founder Evan Malone, subsequent leaders Daniel Cohen and Jeff Lipton, and the undergraduate students who developed the platform over 6 years.

And of course, we thank the current and former students and staff at the Creative Machines lab of Cornell, who created and continue to create the future of 3D printing: Evan Malone, Daniel Cohen, Kian Rasa, Megan Berry, Natasha Gangjee, Adalena Mukergee, Dan Periard, Franz Nigl, Jonathan Hiller, Max Lobovsky, Jonas Neubert, Jeffrey Lipton, Jeremy Blum, Rob MacCurdy, Cheryl Perich, and Apoorva.

We appreciate the hard work and dedication of the team at Wiley, particularly Mary James and Kevin Kent. Mary James came up with the idea of a book on 3D printing and convinced her management that it was a worthwhile topic that people would be interested in. She patiently guided the book through its growing pains. Kevin Kent meticulously helped pull the evolving manuscript into shape and kept us on track for several months running. And of course, thanks to our literary agent Karen Gantz Zahler for seeing the book's potential.

Finally, no book on 3D printing would be possible without the inspiration and genius of the creative and bold innovators out there, the Makers, inventors, artists, and entrepreneurs. Thanks to you, the world benefits from a steady stream of new printing technologies, novel materials, innovative business models, and wonderful new designs and applications. Keep up the great work.

# Contents

# Preface

One of the great things about 3D printing is that the field moves faster than the speed of light and technological advances take place in huge leaps and bounds. Yet, rapid innovation is a difficult topic to capture. Just as you figure out how to pin down an elusive and squirming new idea onto paper, it's already out of date. Everybody has moved on.

We wrote this book over nine feverish months, constantly seeking a broader perspective by speaking with experts we already knew and obsessively combing the web and twittersphere for new information and people to talk with. We chose not to write a simple "here's how to use a 3D printer" since such a book would be obsolete in just a few months. Instead, since the technology changes faster than a book can be produced, we dug below the surface and chose to explore the deeper implications of 3D printing technologies, how this new capacity for production will change our lives, our laws and our economy.

Writing a book with two authors is both a blessing and a curse. There's nothing like having a fellow traveler in the grueling and solitary writing process. No one else in the world is as eagerly willing to discuss and dissect arcane ideas for hours at a time. But having two authors presents a challenge: How do you manage two narrative voices?

Should two authors consolidate their individual experiences and insights into a collective "we?" Or should two authors differentiate each individual thought and experience by the name of the individual author (e.g., Melba or Hod)? After some thought and quite a few revisions, we settled on the following: new insights are "we." Individual past experiences are "I"; however, in the chapters we don't differentiate which "I" that is. So sometimes the "I" is Hod. Sometimes the "I" is Melba.

While writing this book, we had the privilege of interviewing more than 20 experts from different sectors and parts of the world. Their excitement for the topic was contagious and their feedback on the chapters invaluable. Here we'd like to express our appreciation for their time, patience, and best of all, the fact that their creative energy is changing the world as we know it.

We hope you will enjoy the book.

# Fabricated

# 1 Everything is becoming science fiction

Place: Your life

Time: A few decades from now

*. . . even in the future, it's still hard to get up in the morning.*

*The smell of freshly baked whole wheat blueberry muffins wafts from the kitchen food printer. The cartridges to make these organic, low-sugar muffins were marketed as a luxury series. The recipes were downloaded from different featured artisan bakers from famous restaurants and resorts.*

*The first time you showed the food printer to your grandfather, he thought it was an automated bread machine—an appliance from the 1980s that took foodie kitchens by storm. He couldn't understand why you wanted to print processed food until his anniversary came. To celebrate, you splurged on deluxe food cartridges and printed him and your grandmother a celebratory dinner of fresh tuna steaks, couscous and a wildly swirled chocolate-mocha-raspberry cream cake with a different picture within every slice.*

*Managing your diabetes has gotten easier since the health insurance company upgraded your food printer to a high-grade medical model. New medical-grade food printers for diabetics read streams of wireless signals from a tiny skin implant that tracks your blood sugar. When you wake up in the morning, the FoodFabber receives the first reading of the morning and adapts the sugar content and nutritional balance of your digitally cooked breakfast accordingly.*

*After breakfast, it's time to check the news. The top story is an update on a rescue operation of several miners who have been trapped underground for a week. Their mine shaft collapsed, stranding them deep underground. At first rescue teams tried to dig them out until their shovels nearly triggered a deadly rubble slide.*

*Fortunately, the mining company followed federal safety regulations and properly equipped its miners with regulation safety gear. 3D safety printers are a standard tool that mining crews carry with them into deep mine shafts. Before they descend into the shaft, technicians make sure each printer has updated design files for every essential machine part that will go down into the mine. The 3D printer goes down with other machinery in case a part breaks and needs to be quickly replaced deep underground.*

*Today's news update on the mining disaster reports that the portable safety printer has become an unlikely hero. For several days, the trapped miners have been conversing with above-ground technicians over a limping wireless connection. Both teams—one above and one underground—are working together to refine the printer's design blueprints for the replacement parts.*

*What should have been a short, standard rescue operation has grown complicated. Just printing a few replacement parts would have been easy. The reason the rescue effort has been delayed is because the design for the broken part keeps buckling after it's installed because of unexpected high levels of humidity inside the mine shaft.*

*The good news is that the situation in the mine looks brighter today. The reporter explains that the third attempt to print the replacement part passed its stress tests under simulated conditions at the mining company's headquarters. Today the miners will print the updated design underground and, if that works, start rebuilding their damaged machine tonight.*

*As you leave your house for work, a crane and a lone construction worker toil silently on an empty lot across the street. Your neighbor's construction project is the talk of the neighborhood. A few weeks ago your neighbor knocked down his old-fashioned wooden house to fabricate a new eco-friendly luxury home.*

*He waves from the mailbox and shows you the marketing brochure. The new home is a luxury model from a company called FoamHome and will be completed in two more weeks. FoamHome's catalog explains that each home's walls are constructed with built-in weather sensors. The roof, when it's laid on top at the very end of the process, will contain solar panels. Walls will be fabricated with electrical wiring and copper pipes already in place.*

*Together you and your neighbor watch the construction crane slowly maneuver a gigantic nozzle over the top of the new foundation. The nozzle simultaneously scans the landscape and adapts the blueprint, as it squeezes out a paste made of a blend of cement and some synthetic building materials. The crew member's job is to make sure no one walks on site during construction. The brains of the outfit is a small computer attached to the construction crane that guides the fabrication process.*

*The neighborhood has been watching the FoamHome project with great interest as the home's walls slowly grow. What was that old joke about the early days of factory automation? "All you need these days to run a factory is a man and a dog. You need the man to feed the dog and the dog to bite the man if he tries to touch anything."*

*So far, the slowly growing house looks gorgeous, its walls curved in organic patterns and soft curves and hollows. Nobody could build a house like this with frame carpentry, no matter how many people worked on the construction crew. No one has yet seen the inside of the new home, but rumor has it that your neighbor ordered designer inner walls that will look like they're made of old-fashioned brick and mortar.*

*Finally you reach your office and catch up on the details of the final stages of a long investigation you've been leading for months. Your team was assigned to investigate a new sort of black market, one that deals in replacement body parts. More and more patients, desperate for replacement organs, are purchasing them from uncertified rogue bioprinting services rather than a certified medical provider. Bioprinting custom body parts continues to be a controversial topic in the public mind, more polarizing even than the stem cell, abortion, or cloning debates of your grandparents' generation.*

*It's gotten too easy to get replacement organs made. The cost of a high-res full body scan has plummeted in the past few years. People like to get them in their 20s and save the data for later, just in case if something goes wrong and they need a quick replacement organ. Sometimes it's their joints that fail. In reality, the most common use of "body design files" is for cosmetic surgery, to recapture the tight wrinkle-free skin and body of youth.*

*Bioprinting isn't the problem. In fact, most people believe that bioprinting is a life-saving technology. The challenge is what to do about the growth of these new black markets. Regulating the production of new printed body parts is difficult since the cost of bioprinters has also plummeted. Black marketeers snap up cast-off medical bioprinters for less than the price of a new car as last year's bioprinter models are sold off each year by hospitals and surgical clinics.*

*During the investigation, you've learned that most of the time black market organs actually work pretty well. The problems arise from faulty design files or sloppy organ makers who cut corners and don't use a sterile printing environment. In a recent case a few patients died from uncertified "vanity organs" they purchased to improve their athletic ability and appearance. Their families are trying to figure out who to sue: the rogue manufacturer, the bio-ink supplier, the organ designer, or the company that certified the design.*

*Black market bioprinters range from well-intended, would-be healers to deadly, profit-driven peddlers of rogue, counterfeit flesh. Some call the competent and hygienic black marketeers heroes for helping ill people obtain vital new organs at a lower price. Others deplore the organ merchants' eagerness to profit from buying and selling essential tissue to vulnerable people, especially in cases where the new printed organ is poorly crafted.*

*At the end of the work day you stop by your daughter's middle school. You're one of the parental sponsors of this year's Science Fair. Your daughter's teacher tells you that 3D printers are disrupting the culture of the Science Fair. Lazy students 3D print elaborate objects with little effort and no skill—they just need to have a good design file. Many lower income students do not have home 3D printers so they aren't getting the design time and practice they need for a level playing field.*

*There's another twist. The teacher explains that for this year's fair, parents will serve on a clean-up crew. Last year after the fair ended, the school's custodians complained that the gym floor was littered with the debris of dozens of frenzied printing demonstrations. Even worse, for several days after the fair, students and teachers stumbled over dozens of mouse-sized, ready-made robots that clanked and rolled around school hallways. Some printed robots recited appropriate and preprogrammed bits of scientific lore. A few of the roaming robots, however, seemed to have mastered a few unauthorized and slightly more colorful bits of wisdom.*

*When you and your daughter get home, your spouse shares good news. His 3D printing manufacturing business just got accepted into an aerospace cloud manufacturing network. Cloud manufacturing is a new way to make things that's starting to replace mass manufacturing. Cloud manufacturing—like cloud computing—is a decentralized and massively parallel model of production. Large companies order parts and services on demand from a vetted network of several small manufacturing businesses that have joined forces to manufacture specialized parts.*

*Cloud manufacturing is catching on quickly in the electronics, medical, and aerospace industries. These companies need complex, highly sophisticated parts, but not in huge batches. Clouds of small manufacturing companies save the big companies money. Cloud networks tend to be located near their clients so there's less long-distance shipping of printed parts. The companies keep designs for product parts in digital inventory and make just one or a few at a time. Cloud manufacturing networks have been a boon for regional economies everywhere, creating local jobs in specialized small manufacturing and services companies.*

*Your spouse's particular cloud consists of small companies that fabricate specialized fuel injector parts for military and commercial airplane manufacturers. To get into this particular network, his business had to demonstrate its manufacturing prowess by 3D printing sample airplane machine parts in a specified time frame. The manufacturing network stress-tested his sample parts and they performed well. After some negotiation on profit margins and manufacturing capacity, his business was admitted into the network.*

*Finally, the day winds down. Your son likes his bedtime routine in which he brushes his teeth and you tell him a story once he's in bed. Tonight you discover that, as usual, his toothbrush has somehow gone missing. He thinks he may have left it at his friend's house yesterday. You could run to the store to buy a new one, but there's an easier way.*

*You boot up your home Fabber and let your son eagerly scroll through several different toothbrush designs. Several different companies sell designs on the Fabber but your son already knows he wants a zBrush—still a bargain at 99 cents. Your son likes the fact that there are several different cartoon figures offered for the toothbrush handle. You authorize his purchase and scan your son's custom measurements—the size of his hand and shape of his open mouth—with a small wand attached to the Fabber.*

*The Fabber starts printing. On its glowing screen a list of design credits scrolls past that resembles those of a movie—from the designer of the toothbrush program to the company that owns the copyrights of the designs for the cartoon figures. The new toothbrush will be ready for use in 15 minutes.*

*As the Fabber prints you tell your son his bedtime story. It's about the old days, one of those "when I was your age" tales. Your son listens skeptically. He has a hard time believing that when you were young, each toothbrush looked alike. If you ordered something from the Internet it took forever—24 hours—until it was delivered to the door.*

*"Wow," he says politely. "Life must have been hard back then."*

# 2 A machine that can make almost anything

What would you make if you had a machine that could make anything?

In England a technician scans the feet and ankles of Olympic sprinters and puts the data into a computer. The computer does a few quick calculations. The technician 3D prints new track shoes that are customized for each athlete's unique body shape and weight, gait and tastes.

On the other side of the world, NASA test-drives a version of its Mars Rover in the Arizona desert. On board the Rover are several custom-made 3D printed metal parts. Many of these parts have complicated shapes made of curves and inner hollows that could not have been manufactured by anything other than a 3D printer.

In Japan, an expectant mother wants to create the ultimate commemoration of her first ultrasound. Her doctor edits her ultrasound image and 3D prints a precise, highly detailed replica of the fetus. The result, an avant-garde 3D printed plastic tribute to the tiny fetus, encased for posterity in a block of hard transparent plastic.

These modest manufacturing miracles are already taking place. In the not-so-distant future, people will 3D print living tissue, nutritionally calibrated food, and ready-made, fully assembled electronic components. This book is about a new way of making things. In the following chapters, we explain 3D printing technologies and design tools in simple language. For readers of a technical bent, a few chapters delve deeper into the details of 3D printing's current and future. We then explore the downstream implications—economic, personal, and environmental.

3D printing opens up new frontiers. Manufacturing and business as usual will be disrupted as regular people gain access to power tools of design and production. Intellectual property law will be brought to its knees.

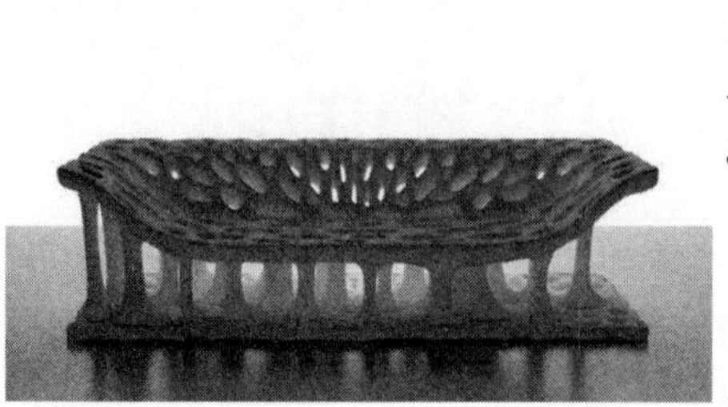

Image courtesy of Andrea Morgante and Enrico Dini, D-Shape

**A printed full-scale bench in stone-like material**

Some people remember exactly where they were when they watched the first moon landing. Others remember the confusing first weeks when the Berlin wall came tumbling down. I remember the first time I heard about 3D printing.

The time was the late 1980s. The place, a tedious engineering seminar on manufacturing engineering. The classroom was warm. The professor had the misfortune to have a droning, unintentionally soothing voice that lulled my classmates and me into a sort of group stupor.

The classroom door banged open, disrupting the peaceful drowsy calm of the afternoon's lecture. An unfamiliar man burst into the room. Our surprise visitor announced he was a salesman from a company called Cubital Systems. We had never heard of Cubital Systems, which at that time was one of the two companies in the world selling commercial 3D printers.

The salesman animatedly waved a plastic object over his unruly mop of hair and announced that a manufacturing revolution was brewing. "I am holding the future of manufacturing in my hands," he boldly claimed. "This plastic object was made by a laser that 'printed' plastic."

Intrigued, my classmates and I stirred curiously and wondered why he had come to our class. At that point, sensing our interest, our professor wisely handed over the classroom to his animated visitor. We learned later that the Cubital salesman had been invited by our professor as a guest lecturer.

A skilled showman, the salesman ceremoniously paused, relishing our confusion. In the quiet that followed, he asked a student to turn the crank sticking out of the plastic object. I can still hear the crisp clicking sound in the classroom as my classmate energetically turned the crank for what seemed like an endless minute. Inside the device, complicated interlocked gears tugged one another into motion.

Our sleepy brains struggled to life and we whispered questions to one another. "Did he just say that he *printed* that thing using a laser?" My classmates and I studied the grinding gears, trying to figure out where this unexpected entertainment was going.

# 'Revolutionary'

## Machine makes 3-D objects from drawings

**By Kathleen Sullivan**
American-Statesman Staff

Wedged into the corner of an unused photo lab at the University of Texas is an ungainly machine that can transform a computer drawing into a three-dimensional model at the touch of a button.

Sometime next year, the machine, which was developed by a UT graduate student, will make its way out of the lab and into the commercial arena. It will leave with the blessing of the UT Board of Regents, which Thursday gave an Austin company exclusive licensing rights to the "revolutionary" new technology embodied in the machine.

The licensing pact paves the way for the first transfer of technology from the University of Texas at Austin to a commercial venture.

The company that won the right to market the invention is Nova Automation Corp., whose principal shareholders are an Austin consulting engineer and Nova Graphics International Corp., an Austin-based computer graphics software firm.

The agreement represents a "hard fought" victory for UT's fledgling Center for Technology Development and Transfer, said Meg Wilson, coordinator of the center, which was given life during the last Texas Legislature and got

See Inventor, A11

Staff photo by Ralph Barrera

Associate Professor Joe Beaman shows some three-dimensional plastic models made by the 'selective laser centering' device developed by Carl Deckard, left.

Image courtesy of Carl Deckard and Joe Beaman

**Carl Deckard and Joe Beaman inventing the first Selective Laser Sintering printer (circa 1986) at the University of Texas**

We became even more confused when the salesman dropped his next bombshell. "All of the gears, handles, and knobs that you see here were never assembled. They were all printed in place as a single, preassembled bunch of distinct parts."

The confused buzz in the classroom increased in volume as our energetic visitor gleefully closed his presentation with a final data point: A computer, not a human, had been the guiding "hand" that steered this miraculous machine through the production process. The salesman rummaged through his bag and pulled out a piece of paper he waved in front of us. The page showed a photograph of a computer on whose screen was a design file for the same plastic object he had just shown us.

The Cubital salesman smiled and asked whether we had questions. Somnolence forgotten, my classmates and I peppered him with questions. What did he mean he "printed" all the plastic parts using a laser? And what sort of manufacturing machine could possibly fabricate something made of interlocked parts that didn't need to be assembled? Could it print in materials other than plastic? And, of course—how much would it cost to get such a machine for oneself?

Traditional manufacturing felt instantly obsolete.

I still remember that day. Our guest lecturer's enthusiastic pitch convinced me that this miraculous machine would indeed spark a revolution in the way we make and design things. I'd never seen such a close connection between the software design of an object (design software was a new and growing passion of mine at that time) and its physical manifestation.

That day was two decades ago. The revolution we were promised didn't happen as quickly as planned. A few years later, Cubital went out of business. Like many pioneering technologies, Cubital's 3D printing process was too complicated and slow, and its machines were too expensive for margin-conscious manufacturing companies to embrace.

I sometimes wonder where that salesman went after Cubital folded. His sales pitch—unabashedly dramatic as it had been—was dead on target. It's just a matter of time until regular people will rip, mix, and burn physical objects as effortlessly as they edit a digital photograph.

## Printing three-dimensional things

Like the magic wand of childhood fairy tales, 3D printing offers us the promise of control over the physical world. 3D printing gives regular people powerful new tools of design and production. People with modest bank accounts will acquire the same design and manufacturing power that was once the private reserve of professional designers and big manufacturing companies.

In a 3D printed future world, people will make what they need, when and where they need it. Yet, technologies are only as good as the people using them. People might fabricate weapons and create unregulated or even toxic new drugs. Our environment may be littered with quickly discarded print-on-demand plastic novelties. Ethical challenges of bioprinting will make stem cell controversy seem simple in comparison. Black marketeers will be tempted to earn quick and dirty profits by making and selling faulty machine parts whose shoddy construction could fail at a critical moment.

When most people first hear about 3D printing, their mind leaps to their old, familiar desktop printer. The biggest difference between an inkjet printer and a 3D printer is one of dimension. A desktop printer prints in two dimensions, spraying colored ink onto flat paper documents. A 3D printer fabricates three-dimensional objects that you can hold in your hand.

3D printers make things by following instructions from a computer and stacking raw material into layers. For most of human history, we've created physical objects by cutting away raw material or using molds to form new shapes.

The technical name for 3D printing is "additive manufacturing," which is actually more descriptive of the actual printing process. 3D printing's unique manufacturing technique enables us to make objects in shapes never before possible.

3D printing is not a new technology. 3D printers have been quietly doing their work in manufacturing machine shops for decades. In the past few years, 3D printing technology has been driven rapidly forward by advances in computing power, new design software, new materials, and the rocket fuel of innovation, the Internet.

**This printer costs about $10,000 USD and is the size of a microwave oven.**

Computers play a critical role in the 3D printing process. Without instructions from a computer, a printer is paralyzed. A 3D printer comes to life when it is fed a well-designed electronic blueprint, or design file, that tells it where to place the raw material. In fact, a 3D printer without an attached computer and a good design file is as useless as an iPod without music.

The way the 3D printing process works is as follows. The 3D printer, guided by instructions in the design file, squirts out or solidifies powdered, molten or liquid material into a specific flat pattern. After the first layer solidifies, the 3D "print head" returns and forms another thin layer on top of the first one. When the second layer solidifies, the print head returns yet again and deposits another thin layer on top of that. Eventually, the thin layers build up and a three-dimensional object forms.

3D printers don't cut or mold things into shape the way humans or traditional manufacturing machines do. Making objects in layers opens up the ability to physically output a broader range of digital concepts. If a shape's design has precise internal hollows or interlocked parts, a 3D printer is the first output device that can realize such designs in the physical world.

3D printed parts and products are creeping into everyday life. Your car's dashboard was designed with the help of 3D printed prototypes to make sure all the various parts fit snugly together. If you wear a custom hearing aid, odds are good that it was 3D printed using optical scan data that captured the precise shape of your inner ear.

Dental labs print custom crowns in less than an hour from X-rays. Printed titanium and ceramic replacement knees are walking around in bodies all over the world. If you've had the good fortune to fly in Boeing's new premium airplane, the 787 Dreamliner, you've placed your life into the hands of at least thirty-two different 3D printed parts.

The secret to 3D printing could be summed up as follows: 3D printers are more accurate and versatile than any other mode of production—be it a human or machine—at fabricating a complex design into a physical object, combining raw materials in ways that were once impossible.

Today, the average home 3D printer can make a plastic object as large as a shoebox. Industrial-scale 3D printers can fabricate an object as large as a car or as small as the barely visible head of a pin. Some people have rigged up custom 3D printers that can print large concrete structures the size of a small house. Other researchers have printed at the micro-scale level, making objects whose details are barely visible by the naked eye.

## On being digital (and analog)

In the mid-1990s, ecommerce and digital media were in their infancy. In his wonderfully prescient 1995 bestseller *Being Digital*, Nicolas Negroponte predicted the demise of "entertainment atoms." Years before the great transition to digital media actually took place, Negroponte correctly predicted that purveyors of entertainment in physical form—traditional book publishers, video rental stores and big television networks—would meet the same fate as the dinosaur.

The demise of centrally controlled mass media and book publishing was just the beginning. The end of the 20th century was about information becoming digital. The 21st century is going to be about bringing the virtual world into closer alignment with the physical one.

The virtual world is a place of freedom where gravity is optional. In a video game, characters can leap over buildings, grow themselves a new arm, and morph into different physical shapes. The virtual world is easy to edit and revise. It's impossible to change the color of the bark on a real tree, but simple to edit its image in a digital photograph. The behavior of the virtual world can be programmed. If the details of a physical object are captured in a design file, the digital "raw material" of the design is modular, made of tiny discrete on-screen bits of light, or pixels.

3D printing technologies will close the gulf that divides the virtual and physical worlds. Of course, a skeptic would quickly point out that the digital and physical worlds already intersect at several points. After all, design and manufacturing processes have been driven by computers for decades. Mass production these days is nearly fully automated (except for the last step—the human-intensive assembly line).

The convergence of the virtual and physical worlds will be a slow and subtle process. It will happen in phases. First we will gain control over the shape of physical things. Then we will gain new levels of control over their composition, the materials they're made of. Finally, we will gain control over the behavior of physical things.

### *Control over shape*

A 3D printer can carefully interpret a digital design file, bringing us one step closer to tapping into the rich creativity and freedom of the virtual world. If you watch an animated movie, it's clear that the scene on-screen was created on a computer. Dinosaurs roam through modern subway stations. Flying robotic soldiers shaped like buzzards shoot deadly lasers at whatever crosses their path.

If a movie switches back and forth between animation and reality, to the viewer it's immediately apparent that there's a clear demarcation between the rich computer-rendered world that's the product of several fertile imaginations... and real life. One way to appreciate the promise and peril of 3D printing is to ponder the tyranny and rigid rules that govern the material world. Since 3D printers build objects in layers, they can make shapes that were once possible only in nature. Curves, hollows, and complicated inside chambers become possible.

The challenge is that atoms come together in unpredictable ways. A digital design can look stunning on the computer screen, but literally collapse when manufactured, unable to survive the discipline of gravity and materials limitations. In contrast, the digital world offers our imaginations intoxicating flexibility and creative freedom. The digital world eagerly embraces forms that are impossible in real life.

### *Control over composition*

In the second stage of convergence, 3D printing will give us precise control over what things are made of, or their material composition. Multi-material 3D printers will open the door to the production of novel objects. This new class of object will be made of precise blends of raw materials whose combined whole will be greater than the sum of its parts.

Imagine a water color kit where blue can be mixed with yellow to form a nearly infinite number of different shades of green. In nature, 22 amino acids combine in different ways to create proteins of staggering variety. A multi-material 3D printer armed with precise instructions from a design file will be able to blend familiar raw materials into novel combinations.

When 3D printing technology evolves, we will see the fabrication of objects made of currently unfeasible blends of materials. We will see machine parts that can heal from failure. Or mesh that can stretch to nearly ten times its original length. Medical devices will respond to a particular patient's blood type or detect changes in temperature.

The second avenue of control over composition lies in a slightly different direction. 3D printers will someday fabricate controllable materials. In the virtual world, all information, no matter how complex, ultimately boils down to its bare essence, two base units: a 1 or a zero.

In contrast, physical things are made of rich, non-modular swirls of raw material whose base units are atoms that are unruly and hard to control. Because of the material diversity found in the physical world, "analog" materials are difficult to capture in digital form in a meaningful way. As a result, analog materials are difficult to precisely copy, control, and program.

Incompatible atoms are a manufacturer's nightmare. True, a 3D printer can't smash open atoms to make them more malleable. What a 3D printer can do, however, is to artfully blend together once-incompatible raw materials into a single printed object.

Electronic circuits are notorious for the fact that their metal parts must first be made separately from their ceramic and plastic parts and assembled later. The fact that the raw materials that make up a circuit's critical components must be made on separate manufacturing machines has dictated that circuit boards be flat and made up of several thin layers.

If the components of electronic circuits didn't suffer from the curse of incompatibility, we could create circuits of all shapes and forms. If we could combine conductive and non-conductive materials together on a 3D printer, we could co-fabricate circuits of all shapes and sizes. We could print mechanical devices whose circuits are already built-in, whose complexity rivals that of the biological world.

Image courtesy of Objet Inc.

**Multi-material 3D printing is in its early stages. This toy is actually a sophisticated engineering project made of several different raw materials that were blended together during the printing process.**

Another way to gain control over the material composition of objects is to voxelize them. A voxel is the physical equivalent of a pixel. Voxels could be tiny, discrete pieces of a solid material. Or voxels could be tiny containers that hold whatever you put into them.

We're just learning to 3D print objects made of voxels. Objects made of voxels offer an alternative to the analog materials that comprise most physical things. If you can make something from voxels, you're one step closer to making it behave more like a programmable object, to controlling its behavior. Control over material composition of physical objects opens the door to the next stage, control over the behavior of physical objects.

### *Control over behavior*

Consider a wooden kitchen table. If you were to use an optical scanner and scan its outer surfaces, you could turn the scan data into a design file. Once the table's physical dimensions successfully made the leap into digital format, it would be easy to temporarily gain full control over the table's design by using design software.

You could edit the table's design file, then 3D print out a new version of the table. However, unless you printed the new table in billions of tiny voxels, the new table would be analog. Its materials, its parts and pieces, would remain innate, unintelligent, continuous, and passive. If you could 3D print the new table in voxels, however, a world of new possibilities would emerge.

As electronic components continue to shrink in size and increase in computing power, someday we will be able to 3D print voxels containing tiny circuits. Like graphic pixels whose perfect merged union creates a beautiful, high-resolution digital image, a perfect union of voxels would create intelligent, three-dimensional active physical objects.

Voxels give birth to intelligent and active raw materials. Instead of 3D printing passive parts as we do today, in the future, we will print active systems, for example, a working cell phone. 3D printers would create smart fabric, ready-made robot life forms, and machines that learn, respond, and think. We will print physical things that contain the intelligence of digital things.

Someday 3D printing will bring artificial intelligence from the computer into the real world. Robots are old hat. Cyborgs are a cultural relic from the 1990s. The future lies in programmable matter, raw materials whose behavior we can program and 3D print in a chosen shape.

MIT professor Neil Gershenfeld, in his book *When Things Start to Think*, predicted that programmable matter will contain a mind of its own.[1] When things start to think, digital processing power will literally find legs and walk into the physical world. 3D printed programmable materials will form their own physical body, complete with mechanical and tactile capacities.

Perhaps one day, 3D printed robotic life forms will emerge from the printer complete with batteries, sensors, and circuit-brain already inside. New-born 3D printed robots will take their first hesitant baby steps out of the print bed and put their electronic circuitry to work learning their way around. Perhaps someday 3D printed robots will return to their 3D printer to invent new features for their "birth machine," for a health check, to recalibrate or replace printed parts.

## Faxing things

The ultimate convergence will arrive when we effortlessly shape-shift between being physical and being virtual, when physical objects smoothly transition from bits to atoms and atoms to bits. In the same way an online document can be printed on paper, scanned, and then printed again, someday physical things will migrate between bits and atoms and back again.

In *Being Digital*, Negroponte cautioned that the physical world won't lend itself easily to digital format. Atoms are heavy and expensive to ship. Physical inventories take up space. Atoms insist on stubbornly clinging to one another in strictly defined ways.

Negroponte wrote, "If you make cashmere sweaters or Chinese food, it will be a long time before we can convert them to bits. 'Beam me up, Scotty,' is a wonderful dream but not likely to come true for several centuries. Until then you will have to rely on FedEx, bicycles, and sneakers to get your atoms from one place to another."[2]

3D printers may someday be the ultimate fax machines. If the virtual and physical worlds were to truly become non-exclusive, we could effortlessly fax things from place to place. Years ago, when I was a graduate student, a few of my colleagues were working hard to figure out how to do this. They were developing optical scan technologies and testing out their accuracy on a primitive 3D printer.

I admired their vision, which even today is still decades ahead of its time. However, as far as I know, they were never able to overcome a few core challenges. First, an optical scan captures only the surface details of an object. Most objects contain important internal structure. Second, my former colleagues could only print simple inanimate objects made of a single material.

Today we can "fax" simple physical objects. I worked on a project with a colleague of mine, an archeology professor. He deciphers ancient cuneiforms from several thousand years ago. His research regularly takes him to archeological sites in various countries in the Middle East.

Recently my colleague returned from a trip abroad, perturbed that he could not bring invaluable information home with him. Ancient cuneiforms are valuable objects whose finders promptly submit them to the governments of the country they're digging in. Archeologists attempt to capture what they can by taking pictures and painstakingly scribing the shape of the characters. The challenge, however, is that there's nothing like the real thing.

My colleague and I decided to do an experiment to see whether we could "fax" priceless cuneiforms from one place to another. We agreed that we would CT scan a few of the cuneiforms he had on hand in his own collection. Then we would turn the scan data into a design file and re-create exact replicas of the cuneiforms on a 3D printer.

If our experiment worked, we figured that the next time he went abroad, my colleague could convince the local authorities to let him CT scan priceless local artifacts. Then he could send the design information to a nearby (or distant) 3D printer to share with anybody else on the planet. He could offer his host country an invaluable preservation service as well, since they, too, could store the data from cuneiform CT scans and 3D print replicas.

Image courtesy of Cornell University. Curator David I. Owen; Design: Natasha Gangjee; Photo by Jason Koski

**CT scanned priceless artifacts can be 3D printed for preservation and educational purposes. On the left is the original cuneiform and its 3D printed replica on the right. An enlarged image of the replica is below.**

We discovered that "faxing" cuneiforms was surprisingly easy. First we converted the CT scan data into a design file. Then we 3D printed exact replicas of rare and precious cuneiforms in different sizes and from different materials.

Best of all, in the process we discovered an unexpected bonus in this cuneiform fax experiment: the CT scan captured written characters on both the insides and outside of the cuneiform. Researchers have known for centuries that many cuneiform bear written messages in their hollow insides. However until now, the only way to see the inner message has been to shatter (hence destroy) the cuneiform. One of the benefits of CT scanning and 3D printing a replica of a cuneiform is that you can cheerfully smash the printed replica to pieces to read what's written on the inside.

## The ten principles of 3D printing

Predicting the future is a crapshoot. When we were writing this book and interviewing people about 3D printing, we discovered that a few underlying "rules" kept coming up. People from a broad and diverse array of industries and backgrounds and levels of expertise described similar ways that 3D printing helped them get past key cost, time and complexity barriers.

We have summarized what we learned. Here are ten principles of 3D printing we hope will help people and businesses take full advantage of 3D printing technologies.

**Principle one: Manufacturing complexity is free.** In traditional manufacturing, the more complicated an object's shape, the more it costs to make. On a 3D printer, complexity costs the same as simplicity. Fabricating an ornate and complicated shape does not require more time, skill, or cost than printing a simple block. Free complexity will disrupt traditional pricing models and change how we calculate the cost of manufacturing things.

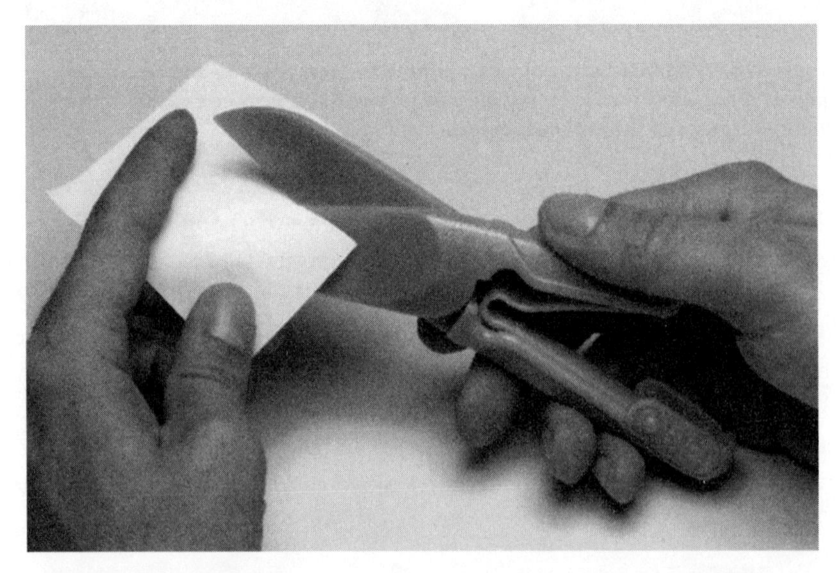

**Printing functional objects. These 3D-printed scissors work "out of the box"—no assembly or sharpening required.**

**Principle two: Variety is free.** A single 3D printer can make many shapes. Like a human artisan, a 3D printer can fabricate a different shape each time. Traditional manufacturing machines are much less versatile and can only make things in a limited spectrum of shapes. 3D printing removes the overhead costs associated with re-training human machinists or re-tooling factory machines. A single 3D printer needs only a different digital blueprint and a fresh batch of raw material.

**Principle three: No assembly required.** 3D printing forms interlocked parts. Mass manufacturing is built on the backbone of the assembly line. In modern factories, machines make identical objects that are later assembled by robots or human workers, sometimes continents away. The more parts a product contains, the longer it takes to assemble and the more expensive it becomes to make. By making objects in layers, a 3D printer could print a door and attached interlocking hinges at the same time, no assembly required. Less assembly will shorten supply chains, saving money on labor and transportation; shorter supply chains will be less polluting.

Image of vehicle printed on ZPrinter 650 courtesy of 3D Systems

**The plastic parts in this image look like assembled bricks but were actually 3D printed, pre-assembled, in a single print job.**

**Principle four: Zero lead time**. A 3D printer can print on demand when an object is needed. The capacity for on-the-spot manufacturing reduces the need for companies to stockpile physical inventory. New types of business services become possible as 3D printers enable a business to make specialty—or custom—objects on demand in response to customer orders. Zero-lead-time manufacturing could minimize the cost of long-distance shipping if printed goods are made when they are needed and near where they are needed.

**Principle five: Unlimited design space**. Traditional manufacturing technologies and human artisans can make only a finite repertoire of shapes. Our capacity to form shapes is limited by the tools available to us. For example, a traditional wood lathe can make only round objects. A mill can make only parts that can be accessed with a milling tool. A molding machine can make only shapes that can be poured into and then extracted from a mold. A 3D printer removes these barriers, opening up vast new design spaces. A printer can fabricate shapes that until now have been possible only in nature.

**Principle six: Zero skill manufacturing**. Traditional artisans train as apprentices for years to gain the skills they needed. Mass production and computer-guided manufacturing machines diminish the need for skilled production. However traditional manufacturing machines still demand a skilled expert to adjust and calibrate them. A 3D printer gets most of its guidance from a design file. To make an object of equal complexity, a 3D printer requires less operator skill than does an injection molding machine. Unskilled manufacturing opens up new business models and could offer new modes of production for people in remote environments or extreme circumstances.

**Principle seven: Compact, portable manufacturing.** Per volume of production space, a 3D printer has more manufacturing capacity than a traditional manufacturing machine. For example, an injection molding machine can only make objects significantly smaller than itself. In contrast, a 3D printer can fabricate objects as large as its print bed. If a 3D printer is arranged so its printing apparatus can move freely, a 3D printer can fabricate objects larger than itself. A high production capacity per square foot makes 3D printers ideal for home use or office use since they offer a small physical footprint.

**Principle eight: Less waste by-product**. 3D printers that work in metal create less waste by-product than do traditional metal manufacturing techniques. Machining metal is highly wasteful as an estimated 90 percent of the original metal gets ground off and ends up on the factory floor. 3D printing is more wasteless for metal manufacturing. As printing materials improve, "Net shape" manufacturing could be a greener way to make things.

**Principle nine: Infinite shades of materials.** Combining different raw materials into a single product is difficult using today's manufacturing machines. Since traditional manufacturing machines carve, cut, or mold things into shape, these processes can't easily blend together different raw materials. As multi-material 3D printing develops, we will gain the capacity to blend and mix different raw materials. New previously inaccessible blends of raw material offer us a much larger, mostly unexplored palette of materials with novel properties or useful types of behaviors.

**Principle ten: Precise physical replication.** A digital music file can be endlessly copied with no loss of audio quality. In the future, 3D printing will extend this digital precision to the world of physical objects. Scanning technology and 3D printing will together introduce high resolution shapeshifting between the physical and digital worlds. We will scan, edit, and duplicate physical objects to create exact replicas or to improve on the original.

Image courtesy of Kerrie Luft

**The titanium heel of this shoe was 3D printed in a single piece.**

Some of these principles already hold true today. Others will come true in the next decade or two (or three). By removing familiar, time-honored manufacturing constraints, 3D printing sets the stage for a cascade of downstream innovation. In the following chapters we explore how 3D printing technologies will change the ways we work, eat, heal, learn, create and play. Let's begin with a visit to the world of manufacturing and design, where 3D printing technologies ease the tyranny of economies of scale.

# 3 Nimble manufacturing: Good, fast, and cheap

At my auto mechanic there's a sign taped to the cash register that says "You can have it done good, fast, or cheap. Pick any two." This enduring joke sums it up. But what if my auto mechanic is wrong?

Halfway across the world, a distant toothbrush factory sells children's toothbrushes for 10 cents apiece. This factory employs advanced machines and technicians. That sounds good, like a high quality outfit. The factory's daily output of toothbrushes is over 300,000 pieces a day. That's fast.

Could it be that my salty old car mechanic got it wrong after all? Maybe you can have it done good, fast, and cheap, no compromises necessary. But maybe he wasn't wrong. There's more to the story than at first meets the eye.

Mass production is riddled with hidden costs and delays. If you consider the big picture from the perspective of a company rather than the consumer, mass production is not cheap. Nor is the process of transforming a design concept into a mass-produced product particularly fast. If you desire anything other than a product aimed at the lowest common denominator, mass-produced products are not particularly good either.

Small-batch mass manufacturing is an oxymoron. The lower the product's end price to the consumer, the more critical are high-volume sales to the company. This is why the toothbrush factory—to earn back the investment involved in producing a simple plastic toothbrush—sells its mass-produced wares in big batches. The size of a minimum order is a whopping 28,000 toothbrushes.

Another hidden cost lies in the product design process. Factory production is merciless. Translating a design concept into a profitable mass-produced product is similar to the sacrifices made in adapting a complex novel into a blockbuster movie. Even a simple design for a plastic toothbrush (which looked fine on the computer) will behave in unpredictable ways when subject to the rigors of an industrial-strength plastic molding process. The laws of economics are similarly ruthless. New design concepts are subject to forced trade-offs between product quality and cost to manufacture.

A factory and an assembly line are an ideal way to cheaply make identical products in high volumes. However, such efficiency is not cheap. Behind even a simple plastic mass-produced product lies an invisible cast of thousands. Companies must invest in skilled technicians and engineers to properly transform a design into reality, to calibrate factory machines and oversee assembly lines.

At the heart of the dilemma of good, fast, or cheap lies a basic rule of economics called economies of scale. Economies of scale are the invisible foundation that supports our modern industrial economy. Economies of scale are what make mass-produced products profitable.

Economies of scale drive down the consumer price of mass-produced products and increase profits for a company. However, to earn back the upfront investments in design and production, companies must sell large volumes of the same product. Only after a significant number of identical products are sold does a company begin to profit from its initial investment.

One of the biggest hidden costs of mass production is the sacrifice of variety. To enjoy the benefits of economies of scale, a company must resist the temptation to change a product's design unless there's a large enough market opportunity to justify the investment. Each design change, each minor upgrade or variation hits the bottom line.

Image courtesy of KenPlas Inc.

**This is an injection molding machine in a factory. Injection molding machines fabricate plastic parts quickly and cheaply but making the mold requires upfront investment and involves design challenges.**

Mass production can't offer both companies and consumers good, fast and cheap goods. How about artisan production? Skilled artisans make custom products in small batches—no assembly line or investments in a factory floor needed. Artisans don't hire teams of skilled designers, engineers and technicians. If a design concept proves to be fatally flawed after it's rendered in physical form, an artisan may have to absorb the cost of wasted materials. However, since only one faulty product was created, an artisan can quickly edit the original design without having to deal with the financial consequences of throwing away hundreds (or thousands) of mistakes. The downside of artisan production is that it doesn't scale.

## Somewhere between mass production and the local farmer's market

Mass production is efficient, increasing company profits and lowering consumer prices. Yet economies of scale take a toll on product variety and customization. In contrast, artisan production handles variety and customization with ease, yet output is confined to small batches. 3D printing technologies offer a new path forward by blending aspects of mass and artisan production.

When the platypus was first discovered, explorers thought it was a hoax, that a prankster had somehow stitched together a furry animal with a duck's bill, webbed feet and a kangaroo's pouch. 3D printing is the platypus of the manufacturing world, combining the digital precision and repeatability of a factory floor with an artisan's design freedom.

Like a factory machine, a 3D printer is automated. A digital design file guides it through its paces. The design file succinctly captures instructions for the manufacture of a particular product. This knowledge can be saved or emailed anywhere.

Like a human artisan, a 3D printer is versatile. A printer can fabricate a broad variety of different types of objects without incurring significant upfront investment. On a 3D printer it costs the same amount of money to 3D print 1,000 unique products or 1,000 identical products. The cost of customization nearly disappears.

Despite these advantages, 3D printed manufacturing offers no economies of scale. Like any extreme personality characteristic, the fact that 3D printing fails to provide economies of scale is both its biggest weakness, but also its biggest strength. Harnessing economies of scale is critical to a company whose business model is based on selling large volumes of a commodity product that earns razor-thin margins. However, if a company's business model is based on selling small numbers of unique, constantly changing or custom-made high margin products, 3D printed production (like the platypus) represents an evolutionary leap forward.

3D printing and design technologies make design and manufacturing more nimble. Small companies have access to powerful tools that were once available only to global corporations. Resourceful businesses, armed with a 3D printer and design software, can provide skilled services of a caliber that were once the exclusive domain of corporate in-house design and engineering departments.

While doing research for this book, we learned that 3D printing is taking root in former manufacturing regions, in the economically decaying rust belt of upstate New York and parts of the Midwest. In these businesses, in many cases, employees are alumni of now defunct local manufacturing plants. When their jobs dried up under the dual assault of outsourced production and factory automation, laid off employees faced a difficult decision: should they move away to where the jobs are, or should they find a way to stay put and somehow make it work?

In the past, no small business could have afforded to buy an industrial-scale 3D printer and pay for enough computing power to run industrial design software. That's changing. One small business we visited was founded by a man whose career spanned the glory days of manufacturing in the western world. We'll call him "Mike" (not his real name since his family preferred that their small business not be identified). Mike's business provides design engineering, prototyping and 3D printing services to regional companies.

We arranged a visit with Mike to learn more. Driving through the Rust Belt was a sobering experience. On the way to visit Mike's company, when we turned off the interstate highway, our cell phone service disappeared. Local people had moved away, leaving behind half-empty towns and cities staggering under a shrinking tax base.

Outside the car window, the picturesque rural landscape had rolling hills, shabby red barns and grazing black and white dairy cows. The two-lane road periodically cut through struggling small towns with disappearing populations. "What do people who live here do to make a living?" was a question that came to mind.

After a few hours of driving we pulled into the driveway of a well-tended home, Mike's company's headquarters. Mike met us at the door and ushered us into his basement workshop and design studio. Part machine shop and part office, the headquarters of this small business looked nothing like the noisy, warehouse-sized production shops of the grand old days when mass manufacturing took place locally. A few computers were tucked away on a desk in the corner of a well-lit room. A pool table sat against the wall, like most pool tables, serving as a handy surface on which to stack boxes and store unused goods.

Mike graduated from high school in the early 1970s. His first employer trained him to be a draftsman. Today it seems hard to believe that manufacturing was once a thriving line of work, a good field for bright and promising young people. Back in those days, however, a person could make a good living by working in the region's manufacturing ecosystem.

Several decades later, Mike was laid off. The company he had worked at for several years offshored its manufacturing and moved its professional staff to a distant part of the country. Nearly 15 years ago, Mike and his family decided they would stay in their hometown, despite the tough job markets of their deflating regional economy.

Mike's company is able to practice nimble manufacturing because today, small services firms like his can afford to purchase their own equipment and software. We asked Mike if we could see the heart of his manufacturing operation, the 3D printer. He led us up a flight of stairs as we carefully stepped around several family dogs that swarmed around our feet. In the family living room, we walked past cats languidly dozing on a kitty tree in a corner. In the garage, humbly awaiting its next assignment was a Stratasys 3D printer, slightly taller and wider than a deluxe refrigerator.

Mike told us the 3D printer is a key piece of equipment that enables his company to add value to client's product development processes. The 3D printer is an efficient and accurate output device that enables designers and engineers to test out design concepts in real life. As Mike put it, "Having my own 3D printer is one of the reasons I can add value as a design engineer."

Mike explained that his job is to offer companies a "very skilled and highly accurate prediction of how a product's parts are going to relate to one another." Having a 3D printer on hand makes that possible. When Mike completes a product design, he first tests it out on his own 3D printer. Once the design is printed, he refines it, if needed. When the design finally meets his exacting standards, he gives the completed design file to his client.

At the end of our visit, we asked Mike whether his work as a design engineer had been transformed by modern computing power and 3D printing. Without hesitation, he said, "Absolutely. A design process for a consumer product that used to take a year now takes 3 months. 3D printing is a huge, huge factor in that. These days, we also make more and more end products for our customers."

## Reducing time from design to product in hand (faster)

As our world speeds up, companies are increasingly eager to shorten their time from design to product in hand. Time to product is a key efficiency metric for companies, meaning that the shorter the time between a design and the functioning end product, the better. 3D printing shortens time to product in-hand by enabling designers and engineers to create on-the-spot product prototypes quickly and cheaply.

A prototype is the rough draft of a product design. Prototypes speed up the process by helping designers, engineers, marketing teams and manufacturers double-check that a design will look, feel and act as planned once becomes physical. A popular service provided by 3D printing companies is making product prototypes for car manufacturers.

Automotive manufacturers save time by 3D printing design concepts and showing them to the project team, even customers sometimes. Mike explained to us how his prototyping process worked. "We're always feeding the car company's marketing department with photographs of our prototypes and physical samples of car parts," he told us. "We send them several 3D printed samples of things that are difficult to estimate from the CAD model."

The day we visited Mike's company, he was working out the details of a prototyping project for a new truck. Unlike the old days when the prototyping work would have been handled by an in-house, full-time design team, this particular car company outsourced their product design work. A regional sales force and a marketing department were all that was left of its staff. A factory produced some car parts that were still cheaper to make domestically rather than offshore.

3D printed any prototype is made are slowly but surely replacing hand-carved models of foam or clay. Mike's client used both methods, hiring artisans to hand-carve custom foam models of car parts, and hiring firms such as Mike's to 3D print prototypes of car parts. Mike told us that his client asked him to scan the first round of foam prototypes and then transform the captured data into a detailed design file. "Today we're going to pack up the scanner and go out where they're designing the truck," he said.

However, before any prototype is made, whether hand-carved or 3D printed, a truck's design begins with its marketing and engineering departments. After several brainstorming meetings in which a new truck's specifications and product goals are clarified, marketers and engineers submit information to artists who create several detailed, realistic-looking concept sketches. "A lot of decisions are made on paper before anything really happens," Mike said.

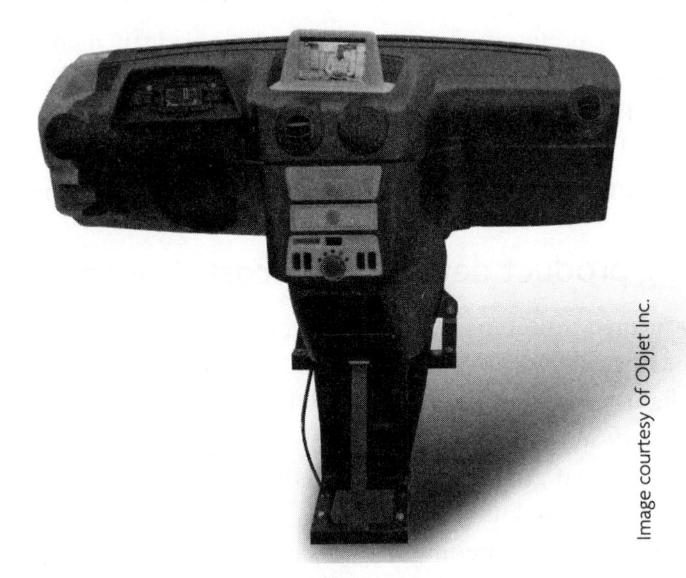

Image courtesy of Objet Inc.

**A 3D printed life-sized prototype of a truck cab, complete with working parts**

Once a new truck's concept sketches are approved, the prototyping process begins. Prototypes give both the marketing team and designers a sense of the truck design's ergonomics and the spatial relationships of its parts. In a truck's cab, a prototype of dashboard knobs, headlights, and other truck parts enables a product manager to experience how the truck will look and feel to a future customer. Prototypes of a truck's engine help mechanical designers make sure that the engine is repairable.

In the old days, when prototyping was still a slow and expensive process, it was risky for a company to cut a corner and just trust that a design would work out in real life. "In the past we ended up with a really pretty design, but then once it was made, we found other issues," Mike said. "For example, if you have a car, you have to repair stuff and do oil changes. There's nothing like trying to reach through a hole in the engine to do an oil change and your hand won't fit."

Someday, perhaps the hand-carved foam models will be a relic of the past. Most companies now skip the foam or clay prototyped models entirely and, instead, jump directly to 3D printed prototypes. If a company uses both hand-carved and 3D printed prototypes, the hand-carved prototype typically comes first. Then, a skilled service provider such as Mike will capture its exact physical dimensions using either an optical scanner or a coordinate scanner.

Hand-carving prototypes is an art. "Some of the old foam models are really things of beauty," Mike said. Then again, 3D printed prototyping can also be an art. Colorful 3D printed models of cars or motorcycles look confusingly realistic. A 3D printer can fabricate complex and unusual shapes that enable designers to dream up futuristic or entirely novel new designs. And, it's faster.

## Reducing product development costs (cheaper)

Some 3D printed prototypes are used by companies to demonstrate design concepts. Other prototypes are used to test out another phase in a product's lifecycle: figuring out how product parts will be mass produced. "Test and fit" 3D printed parts minimize the pitfalls inherent in designing a very complicated product. Despite our growing ease with all things digital, there's no substitute for the simple act of holding physical parts in one's hands.

A test and fit prototype might be a bunch of unassembled parts that engineers attempt to put together in a sort of manufacturing dress rehearsal. When Microsoft surprised the world with by announcing a previously top-secret product concept, a hybrid tablet/laptop called Surface, the media wondered how the company had successfully kept product development under wraps. Usually an early announcement of a radical new tech product is accompanied by rogue photos leaked from a manufacturing facility. The Microsoft hardware division was able to keep product development secret by 3D printing prototypes on machines buried deep in a campus building.

A second goal of test and fit assembly is to make sure that factory machines can physically produce a design concept. In engineering product design courses, students spend several weeks learning about the difficult trade-offs between a designer's great idea and the realities of the factory floor. Thick textbooks describe in painful detail what design ideas will and won't work on a production line. Typical factory machines that mold or cut products into shape have difficulty making an object that is hollow inside, has interlocked parts, or has a complicated internal structure. But not all production challenges can be avoided just by following textbook guidelines.

Upfront investment is squandered when a company discovers too late that its new product's parts don't fit together. 3D printed test and fit prototypes help cell phone product designers lay out tiny hardware components to fit nicely inside a device's sleek case. Hearing aids, car dashboards, razors, hair combs, and smartphones must feel good to the touch and fit comfortably with the human body. Design software and computer simulations are getting better and better, yet even the best designs don't always work out exactly as planned once they're in physical form.

## Making the best custom parts (better)

One of the fastest-growing applications of 3D printing is custom, end-use parts. Custom parts aren't prototypes. They're the real thing. If you prowl community forums, people who have their own 3D printer at home swap advice and design files for replacement parts for standard household appliances from doorknobs to shower curtain rings. Some people 3D print knobs, gears or other antique or discontinued parts that would be extremely expensive to make by hand.

Since custom parts don't benefit price-wise from economies of scale, small, skilled 3D printing service providers are finding new business opportunities. Car and motorcycle companies—even the company that created the Mars Rover—use 3D printed custom parts to build working versions of a concept vehicle or machine. After all, it's critical that million-dollar vehicles be taken for a real test-drive.

The medical industry and dental industries are heavy consumers of 3D printed end parts since their products must interface closely and precisely with the body. Custom-fit dental braces and crowns, previously custom-made, are increasingly 3D printed. Hearing aids and prosthetic limbs are printed from data scans of a patient's ear canal or remaining mirrored limb.

Aerospace companies use 3D printed custom parts in commercial airplanes. Next time you fly in a new airplane, check out the adjustable air duct over your seat. An airplane air duct is a perfect example of a highly custom and costly part that does not benefit from an economy of scale. Especially since unlike toothbrushes, only a few new airplanes are made each year.

When we visited another small service provider in a rust belt in the Midwest, he showed us a semi-translucent, plastic 3D printed object he had been tinkering with. The object looked vaguely familiar but difficult to precisely identify. It was about the size of a garlic press and was made of several round, interlocked sockets and gears.

We couldn't place the odd looking gadget until its designer enlightened us. "This is a 3D printed air vent for a 747," we were told. Aerospace manufacturers are eager to cut down on part assembly. The parts of a 3D printed air vent can be fabricated in a single, already assembled piece.

## Sizing the market

It's impossible to know how many small businesses, exactly, have gravitated towards the marketplace for nimble manufacturing services. However, some robust market data exists. Each year, additive manufacturing industry consultant and analyst Terry Wohlers puts together an annual *Wohlers Report*. The report features market and usage data, plus case studies gathered from roughly 100 of the larger companies worldwide that sell 3D printers, or provide or purchase printing services. The *Wohlers Report* is a goldmine of insight into the additive manufacturing industry. It has become the industry's unofficial "state of the union" assessment for analysts, for executives and journalists alike.

When Terry Wohlers interviews the world's leading 3D printing companies, one of the questions he asks them is who's buying their printers or hiring their printing services. His research shows that consumer electronics companies are heavy users of 3D printed products and services. Next are companies in the auto industry, the medical and dental industries, and aerospace manufacturers.

Wohlers market data indicates that roughly 40 percent of the world's printers are located in the United States. Companies in Germany and Japan are also actively exploring and using 3D printing, each hosting about 10 percent of the world's printers. Finding solid data on the presence of 3D printing in China is difficult. According to Wohlers data, despite its dominance in mass manufacturing, China is home to only about 8.5 percent of the world's 3D printers.

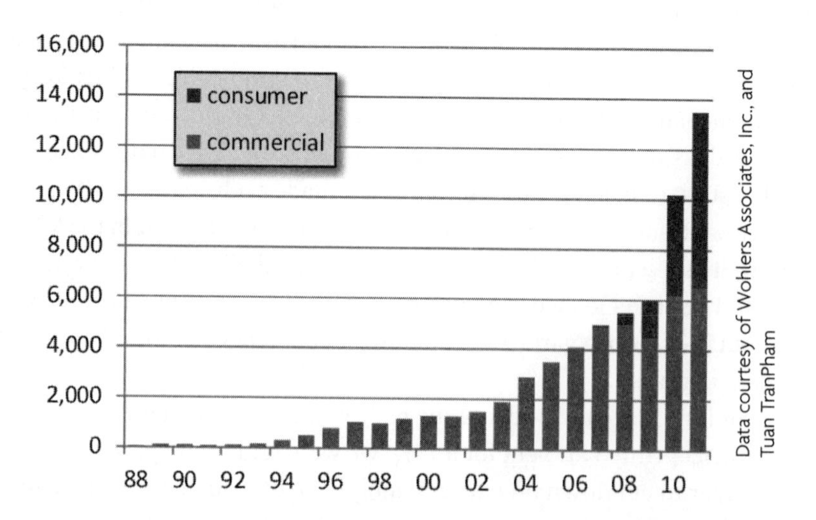


**3D printers sold per year. Sales of consumer-grade machines (less than $5,000) begins in 2007 and outpaces sales of industrial-grade machines in 2011.**

The companies that are the heaviest consumers of 3D printers and services tend to have deep pockets. Yet the size of the market for 3D printed goods and services remains tiny, particularly when the 3D printing market is compared against the size of the $15 trillion a year market for mainstream manufacturing goods and services. Terry Wohlers estimates that in 2011, a total of $1.7 billion U.S. exchanged hands around 3D printing. If the global market for mainstream manufacturing were the volume of a beach ball, the volume of the worldwide market for 3D printing would be comparable to a ping-pong ball.

On the other hand, although a $1.7 billion industry is a modest piece of the world's economy, this market may not be as small as it sounds at first. Small manufacturing-related businesses like Mike's dot the global manufacturing landscape. In fact, the U.S. Census Bureau data indicates that about half of the manufacturing firms in the U.S. employ fewer than ten employees, and a quarter employ fewer than five. A 3D printing industry that's worth a few billion dollars a year may be insignificant in the big picture, but a few billion dollars a year distributed amongst many small businesses could go a long way.

## The blank canvas of the 21st century

Companies that make 3D printers keep a careful eye on the new world of nimble manufacturing. I paid a visit to one of the leading 3D printer companies to get more insight. From the outside, 3D Systems' headquarters looked like any other successful mid-sized technology corporation that's home to more than 700 employees. The building's white and glass exterior did not hint at the creative energy inside.

Visitors to 3D Systems walk into a waiting area that's worthy of an exhibition in a museum of modern art. Clean white walls showcased brightly colored 3D printed objects. To the right side of the room, a sleek red race car the size of an amusement park bumper car sat proudly on a pedestal. On the left, a counter was laden with dozens of colorful 3D printed toys and random machine parts.

Nearby, two 3D printed fully functioning funky plastic electric guitars are displayed. One guitar was purple and the other bright red, designed by Olaf Diegel in New Zealand. The guitars' bodies were a feat of 3D printed product design. To lighten their weight and optimize their sound quality (and to give them a funky look), their bodies were lattice-like structures punctuated with squiggly cutouts and oddly shaped cavities.

Image courtesy of Olaf Diegel, New Zealand

**This printed electric guitar was made on a 3D Systems sPro 140 SLS system.**

3D Systems, based in the United States, is one of the largest and oldest companies that sell 3D printers, the IBM of the 3D printing industry. 3D Systems has been selling high-end 3D printers since the 1980s. The company's business strategy rests on two goals. One, to continue to build and sell powerful, high-end 3D printers to industry. Two, to fulfill the promise of the company tagline, to provide "3D content-to-print solutions" by building a global end-to-end platform to help people bring design ideas into the physical world. 3D

Systems is investing heavily in a belief that people will flock to the first product that makes it easy to bring digital content into the physical world.

Abe Reichental, CEO of 3D Systems, likens 3D printing to the "blank canvas of the 21$^{st}$ century." Abe sees 3D printing as a great equalizer. "My assertion is that everyone can be creative if you remove all the friction and intimidation," he told reporters at the launch of the company's new consumer product, a sleek, gleaming low-cost 3D printer called The Cube.[1] The Cube resembles a high-end home espresso machine and is 3D Systems' first foray into the home-scale markets. Its handsome appearance contrasts with the intentionally home-spun look of other leading home printers such as MakerBot's popular Replicator.

Innovation is a long-standing company tradition. In the mid-1980s, 3D System's founder Chuck Hull invented stereolithography, a printing technique that's used today by many of the world's industrial-grade 3D printers. I noticed during my visit, in the back of the reception area sat a vintage SLA1, a symbol of 3D Systems' long track record in the industrial space. Similar in appearance to an old 1960s mainframe and roughly the size of a vending machine, the SLA1, the world's first commercial 3D printer, rested peacefully, plugged into a now-vintage 1987 IBM PC.

Behind the reception desk the Willy Wonka-esque ambiance continued. Several glass-walled rooms hummed with people conducting company business. A suitably magisterial-looking company boardroom was visible through a wall of glass, complete with large, dignified looking mahogany board table. In another glass-walled room, teams of white-coated technicians bustled around like nurses tending to patients. Inside, two dozen 3D printers of all shapes and sizes—3D Systems' entire printer product line—were represented. The technicians were fine-tuning each printer's performance and testing out the properties of new materials.

Rajeev Kulkarni, the company's Vice President of Consumer Products, explained that to bring consumers into the game, the 3D printing industry needs a platform, one which spans the entire creation lifecycle, from 3D scan, to design, to print. Rajeev said, "People are used to Google. They're used to Microsoft. They're used to Amazon. We're taking the best elements of all these strategies and bringing them together onto a platform that enables them to experience 3D the same way."

3D Systems' platform strategy takes its cue from the world of software, where the consumer experience, as well as a company's market dominance, rest on its ability to build and control a software platform. For example, Microsoft built Windows and then Office into the world's leading platform for office productivity. In the mobile device space, Apple's iPad, iPhone, and iPod offer consumers intuitive, easy-to-use products and that spawned an entirely new way to sell software in the form of apps.

To build its platform, 3D Systems has been on a dizzying buying spree, acquiring roughly two dozen companies over the past 3 years. The acquired companies represent several different facets of the design and production process. Some offer design services, others make printing materials, build printers, or create content and tools, for example, a simple interface that helps users design simple 3D-printable robots.

Rajeev continued, "While 3D Systems is good at making 3D printers, our goal is to bring the entire 3D capture/create/make process to everybody. That cannot be achieved by just handing people a 3D printer. That has to be achieved by presenting them a platform."

Cathy Lewis, Vice President of Marketing at 3D Systems, is a long-time visionary in the arena of low-cost 3D printing. Before she joined 3D Systems in 2010, she was the CEO of Desktop Factory, one of the first companies to build commercial 3D printers aimed at the home and office markets. Desktop Factory made democratic 3D printers that—unlike the mostly industrial 3D printers available at the time—printed nylon using concentrated light from an everyday halogen lamp, not a laser.

When Chuck Hull founded 3D Systems in 1986, it's unlikely he would have predicted that just a few decades later there would be a small but growing market for home-scale 3D printers. When Cathy Lewis first showed off the Cube at electronics trade shows and industry events, she discovered an unexpected degree of interest not just from consumers, but also from small manufacturing companies. "At first we were surprised by the number of people who started asking about whether they could use the Cube for end-use manufacturing," said Cathy. "I think there's a need in the market place to help companies envision how to break through their current paradigms to capture the benefits of this new type of manufacturing."

Today, 3D Systems keeps track of several game-changing trends that will shape the future of the 3D printing business. The company believes that cloud-based computing, Big Data analysis, and the ever-growing presence of powerful and smart mobile devices will open up new creative and business opportunities for everyone. It all comes down to finding useful applications for 3D printers that make people's lives better and easier.

"To solve customer problems, you have to look at application needs," said Cathy. "It's not just about 'let's create a machine that's faster and better.' It's 'let's create a machine that truly solves this problem that exists today.'"

## The search for the killer app

Today consumer-level 3D printing technologies are in the "Altair phase." In the 1970s, the first personal computers like the Altair were clumsy do-it-yourself kits that their technically skilled users put together themselves at home. With a few exceptions, most of the first few thousand low-end printers sold were patiently assembled and involved a lot of calibration and sometimes troubleshooting.

When I spoke with Rajeev, he described one of the biggest barriers to mainstream adoption of 3D printing: the absence of a killer app, or hugely popular application. "Look at the iPad," he said. "The tablet market was struggling before the iPad came in. It was not the most technologically advanced product, but it revolutionized the industry because it was simple, accessible, and easy to use. If we focus on applications rather than technology, we have a much greater chance of getting this technology adopted."

A killer app—for example email, Facebook, and Angry Birds—is a product or tool that creates new markets, new business models and lures customers to a new technology. These particular killer apps attracted millions of new customers to the personal computer, the Internet, and the iPad. To spawn its own killer app, the 3D printing industry needs a user-friendly platform of tools and an application or game whose mass appeal will create new markets and attract millions of new users.

In part because of the absence of a killer app, the average consumer or small business hasn't yet felt compelled to purchase a 3D printer for home or office use. The market for 3D printing still lies in the manufacturing and design industries. If the dollar value of the worldwide market for all 3D printers and services is the volume of a ping-pong ball, the dollar value of the global market for just consumer 3D printers and services would be even smaller, roughly the size of a grain of rice.

Investment advisor the Motley Fool is bullish on the potential of companies involved in the consumer 3D printing space. In contrast, analyst Terry Wohlers is cautious. He warns that although consumers are eagerly paying high prices for beautiful 3D printed custom goods, "most consumers will never own or operate a machine to produce these products. Instead, they will go to Shapeways, Amazon, or to another service or storefront to purchase these products. Most will not know, or even care, how the products were made—no different from the way they now purchase products."[2]

Yet there are hints of growth potential. 3D Systems annual report stated that between 2010 and 2011, revenue from sales of mid- and small-sized personal and professional-grade (not industrial-grade) 3D printers increased about 40 percent from the year before. Industry-leading companies like 3D Systems, MakerBot, and PP3DP are investing heavily to create consumer-friendly platforms and products to entice users. Outside the commercial world, a growing Maker community buys and builds home-scale 3D printers and shares innovative design files and free advice with the world. Small artisan-style businesses can earn a living by designing and selling custom 3D printed custom machine parts or jewelry or *objets d'art*.

Someday, the technological limitations and barriers that discourage every day use will gradually diminish. The key is to make 3D printing technologies more fun, more social, and of course easier to use. Such an approach is reminiscent of Apple's early consumer strategy. A few decades ago, when personal computing was just coming into the mainstream, Steve Jobs explained why regular people liked the Macintosh. "Most people have no concept of how an automatic transmission works, yet they know how to drive a car. You don't have to study physics to understand the laws of motion to drive a car." When the 3D printing world comes up with its own killer app and creates a vibrant user-friendly, end-to-end platform, the market will explode.

## Connecting artisan and mass production

Science fiction writer Poul Anderson was reputed to have said, "I have yet to see any problem, however complicated, which, when you looked at it in the right way, did not become still more complicated." Predicting future markets for 3D printed goods and services is an equally daunting task. It's difficult—no, impossible—to offer a few crisp words that sum up potential new business models that offer good, fast, and cheap products or services to their customers.

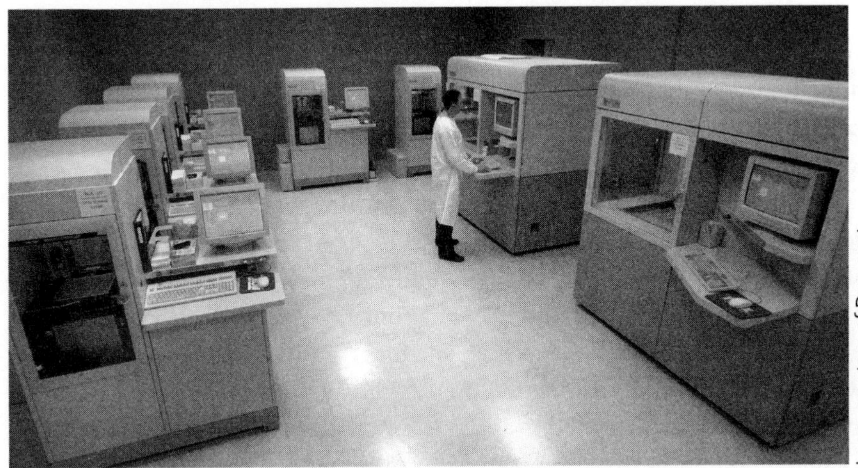

Image courtesy of Dynacept

**Could this be the factory of the future, a small quiet room full of 3D printers?**

For over a century now, most of the world's industrialized nations have been home to two distinct commercial worlds, which like parallel lines run side by side but never meet: mass production and artisan production. Companies seeking new market opportunities should look beyond products and services that fall neatly on one parallel line or another. Opportunity lies in products or services that in the past, because of the limitations of mass and artisan production, have been impractical or impossible to offer at a profit. In other words, opportunity lies in business models where profits are not reliant on economies of scale.

Imagine, for example, if my car mechanic (a great skeptic of the existence of car repair that is good, fast, and cheap) decided to venture into a new line of business: 3D printing car tires. This would be a bold business strategy for him, given that car tires are a classic, mass-produced commodity product. In a corner of the greasy hanger where he fixes cars, he would install a 3D printer that could create precise shapes in high-grade, durable tire rubber.

Out of the gate, let's suppose his first business model would be to make and sell 3D printed tires identical to the brand name tires sitting in his inventory. To anyone familiar with 3D printing, this wouldn't sound wise. But let's assume that my tenacious car mechanic charged forward with his plan anyway.

The results would be disappointing. If he copied the tire designs of the big tire retailers, his 3D printed tires that would remain just as dull and soulless, but be more expensive. Eventually, my mechanic would learn the hard way that economies of scale are unforgiving. 3D printed tires would cost customers more to purchase (not cheap). Custom, 3D printed tires would take longer to make on the spot than just rolling out four new tires from his inventory of mass produced, name-brand tires (not fast). True, the 3D printed tires would be of excellent design and quality, but just quality, alone, can't sustain a business.

Finally, after a near brush with bankruptcy, my car mechanic would sit down and re-think things. His revised business model would no longer rely on the efficiencies of mass production. Instead, his new 3D printed tire business would offer custom-designed tires made for customers willing to pay more.

Compared to ordering a set of custom tires made by a Formula 1 tire specialist, his 3D printed tires would be cheap and produced relatively quickly. Compared to the one-size-fits-all brand name tires in his on-site, inventory, his 3D printed tires could have superior features.

Unlike their bland brethren, custom-printed tires could be made in small batches. These made-to-order tires could perhaps proudly bear a mirrored imprint of their owner's name that would leave personalized tire tracks in sand or snow. Like the market for vanity license plates, some customers would pay more for vanity tires. Imagine the marketing and branding possibilities. An affluent university could sell tires branded with its iconic bell tower to its loyal alumni.

Since, the local market in a small town isn't large enough to provide sufficient walk-in customers to sustain such a business, my mechanic would sell his wares to a global market using a web storefront. Customers could browse tire designs or even upload their own design for a vanity tire. To save on shipping costs, he could strike deals with car mechanics in different locations who would get a commission for printing my mechanic's custom tires locally.

Decoration aside, custom tires could harness the power of 3D printing to improve product quality. Each 3D printed tire could be designed using computer algorithms to afford the best traction tailored to the local climate. Owners of specific car models (or aggressive drivers) could order custom tires to improve their car's performance and reduce their chances of an accident. Some future designers would likely come up with radical new tire designs that would improve a tire's rolling efficiency or wear resistance.

True, to make a profit, this new generation tire company would charge significantly more for the 3D printed custom tires. But if enough customers were willing to pay top dollar for original, branded tires that offered optimal product performance, then selling 3D printed tires could be a pretty good business model. My mechanic would have to edit the old sign taped near his cash register to say "You can have it done good, fast, or cheap. Pick any two. (Unless you're here for the 3D printed custom tires.)"

# 4 Tomorrow's economy of printable products

One of the most commonly asked questions about 3D printing is whether this new technology will create or destroy jobs. Usually I'm asked this by policymakers, local politicians and journalists. After a recent talk on the future of 3D printing at a middle school, I was a little startled when a student raised his hand and asked me, "Will 3D printing help make jobs?"

I gamely explained to the student that there's not a simple answer to this question. If the past is any indication, 3D printing, like other disruptive technologies, will re-shape the occupational landscape in new and unpredictable ways. I asked his classmates whether they had ever heard of a travel agent. A few had. I described how the Internet made travel agents obsolete, but in return, opened up a new market for travel-related services. Similarly, 3D printing technologies will enable new business models while eradicating others. Some jobs will disappear while entirely new professions will emerge.

Fortunately, the student seemed satisfied with my answer, but his question made me wonder what his generation will witness in their lifetimes. Today 3D printing is already becoming a mainstream tool in industries such as aerospace engineering where product lines involve small batches of complex parts. In the future 3D printing will disrupt the economy in more profound ways. Global supply chains will be replaced by agile and independent small manufacturers able to respond quickly to fluctuating inventories and market demands. Less directly, perhaps the biggest contribution of 3D printing technologies to the economy will be to reduce the risk and friction associated with trying out new business models.

## Like ants with factories

One future business model enabled by 3D printing and new design technologies will be cloud manufacturing. Cloud manufacturing, an alternative to mass production, will consist of a network of small-scale, decentralized nodes of production.

We're in the dawn of an era of big, interconnected, yet decentralized internetworked systems. We have Big Data. We have Big Business. Massive financial networks span the globe. We have elaborate economic ecosystems where a monsoon in Thailand impacts the bottom line of a small business in Brooklyn.

Yet mass manufacturing remains a centralized process, concentrated in hubs of specialized activity because it is enabled by economies of scale. Production takes place in factories. Product design is concentrated mostly in professional firms.

Compare the way the telecommunications industry has changed over the years. Once upon a time, phone networks were large, centralized, and unwieldy. A single pay phone would serve several city blocks of people. Each family home had a single landline.

Today, an estimated 60 percent of the world's population owns a cell phone. Each phone is physically tiny, when it stands alone. However, when billions of cell phones all over the world are woven together into a distributed global network, the combined effect is a disruptive, powerful, and gigantic system.

Mass manufacturing still resembles a phone booth rather than a network of cell phones. But this is going to change. Like billions of cell phones, manufacturing might someday consist of millions of small autonomous nodes of production.

3D printing is the catalyst that cloud manufacturing has been waiting for. Cloud manufacturing will be a decentralized system, built on a foundation of ultra-large networks of small manufacturing companies. Wikipedia defines cloud manufacturing as systems "where various manufacturing resources and abilities can be intelligently sensed and connected into wider internet, and automatically managed and controlled."

In his landmark book, *The Long Tail: Why the Future of Business is Selling Less of More*, author Chris Anderson described the collective power of bloggers as that of "ants with megaphones."[1] Until the Internet gave them a worldwide platform, individual writers struggled to make their voices heard. Now, the collective communicative capacity of bloggers exceeds that of journalists working for large media companies. 3D printing technologies will make Makers, consumers, and small companies into ants with factories.

Each individual manufacturing node will be autonomous, yet connected. Manufacturers will form and re-form in temporary collectives as needed around a particular project. Economies of scale will no longer dictate business models since per part, it will cost the same to make one or ten thousand. Each individual company will be versatile, able to make a wide variety of different products or parts on short notice, on demand. Sometimes nodes will pool resources.

Cloud manufacturing will fuel innovation by lowering barriers to entry. Innovation has flourished more quickly in the software industry because the cost of entry is lower in the virtual world than in the physical world. Software products aren't kept in physical inventory. A software product has no raw materials and fewer logistics of transportation and assembly.

Here are two future scenarios.

First, imagine you need 10,000 staplers by the day after tomorrow. You place your design file and order onto the manufacturing cloud. In response, thousands of small companies and individuals, each with a 3D printer (or two) prints out a dozen or so staplers and ships them to you.

Wait a minute. Such a model wouldn't be cost-effective compared to mass manufacturing 10,000 simple staplers. Cloud manufacturing simple commodity items doesn't make sense. It would be cheaper and faster to mass produce 10,000 identical staplers in a few hours in a factory that has raw materials and machines at the ready since it specializes in mass producing staplers.

What if, however, you needed a custom motorcycle made overnight? You would put out a call on the same manufacturing cloud. Thousands of companies would be automatically matched with the task of printing a custom part and shipping it to you when it's completed. The next morning, you would wake up and outside your door would be a small avalanche of a thousand or so custom motorcycle parts, some metal, some plastic, some big, some small.

When you use cloud computing, you don't know where the cloud's computing capacity is located. Cloud-based resources are scalable and automatically load-balanced. When you use the manufacturing cloud you won't have to worry about production capacity. The cloud will intelligently distribute production and coordinate individual companies on the network to successfully absorb your manufacturing request.

You simply place your order and click "Submit."

Each cloud manufacturing company, alone, by itself, may be small. However, like billions of cell phones or ants with factories, the combined whole will be greater than the sum of its parts.

## The Maker movement

Technological disruption happens when regular people get their hands on new tools and apply them to their daily lives. Ubiquity is what enables new technologies to stir up revolution. 3D printing isn't yet a household technology, but it's taking its first steps into the mainstream thanks to a vibrant and growing community of Makers.

The Maker movement is a celebration of do-it-yourself (DIY) innovation. Nobody quite knows where the name "Makers" came from, but it has stuck. The concept of a Maker is similar to what software companies call "power users." Somewhat like the term "hackers" for people who like to bend software to their will, Makers like to bend technology to their will.

O'Reilly Media is a reliable barometer of what people who like technology care about. In 2005, O'Reilly launched *MAKE* magazine to connect to "a growing community of resourceful people who undertake amazing projects in their backyards, basements, and garages. O'Reilly was right on target. *MAKE* has grown into a suite of events and publications that celebrate the do-it-yourself movement. The Maker Faire (The Greatest Show (and Tell) on Earth) has become a gathering place for people who like to make things or like to appreciate what other people have made. Local Hackerspaces and "gyms for inventors" are popping up in cities all over the world.

People involved in the Maker movement tinker in all kinds of technologies, not just 3D printing. Some self-described Makers design circuits, create responsive clothing or musical instruments that glow in response to pitch, or create their own robots. The Maker movement attracts people who crave tools of production, people who have their own tiny CNC routers and milling machines in the basements.

What makes somebody long to spend their Saturday afternoon assembling a 3D printer? Or patiently pinning together tiny electrical components to make their 3D printed toy walk? Lots of reasons. Some people like to get their hands dirty. Others (myself included) find pleasure in solving a concrete problem now rather than having it drag on for months. Joy is a powerful motivating force. I think there's more to it, though.

Makers create wonderful things. Browse around Flickr, or google "3D printing diy" and you'll see printed gargoyles, model train sets, and translucent chess pieces. Wander around a Maker Faire, and you see the ingenuity of what humans are capable of, given time and resources. One exhibit that remains stuck in my mind was The Sashimi Tabernacle Choir that I saw at a Maker Faire in Queens in 2011. The Choir consisted of 250 custom-made electromechanical fish and lobsters that wiggled and sang on the roof of a car.

**A sign directing fairgoers at a Maker Faire 2010 in Queens, New York**

There's not much specific information on the background and motivations of people who describe themselves as Makers. I read a survey—one of the few I've seen—that asked specific demographic questions of people in the 3D printing Maker community. The survey polled 358 people and was created by an organization called Manufacturing in Motion.

Here's what the survey found. Somewhat like software hackers, Makers tend to be affluent males from Europe or North America. Female Makers made up about 7 percent of the number of people that responded. Makers are highly educated; nearly 60 percent of the respondents had at least a bachelor's degree.

The survey asked a key question: Why do you like to make things? More than 80 percent of the people who answered said because they enjoyed 3D printing, and it introduced them to other people who also like to build things. 98 percent of respondents said that creating things with electronics, software, and 3D printers was just good old-fashioned fun.

One of the great things about the Maker movement is the fact that profit is not a core incentive. Makers can be playful and take creative risks since they aren't beholden to massive supply chains, thousands of employees, angry shareholders and other sobering responsibilities that professional designers and manufacturers must grapple with. The core ethos of the Maker movement is community, creativity, social change, and problem solving.

The Maker movement itself may not yet embody the first stirrings of a full-blown, industrial revolution. However, Makers play a critical role in propelling 3D printing technologies into mainstream awareness. Makers, like other early adopters of disruptive technologies, demonstrate what may be someday possible on a larger scale.

It's tempting to compare 3D printing to the early personal computing movement of the 1970s or a new industrial revolution. I've made both of these comparisons at least a few times. These analogies are seductive because it's difficult to concisely describe the sweeping social effects that will be wrought by 3D printing technologies.

Consider the parallels. 3D printing technologies, like mainframe computers, got their start in industry. The first personal computing kits were primitive, low-cost and involved home assembly. The people who first embraced personal computers were similar demographically to the people who have embraced home-scale 3D printers.

There's another complicating factor that increases the allure of personal computing and industrial revolution metaphors: 3D printing is more than a single technology. It's a broad platform technology that will drag along other technologies in its wake. Similarly transformative technologies like the steam engine or telegraph also sent shock waves in every which direction.

## The experience economy

In their book *The Experience Economy,* authors Joseph Pine and James Gilmore predict that a company's competitive advantage will be increasingly based on the intensity of customer experience. Pine and Gilmore explain that the economy has evolved several times already, from an agrarian economy, to an industrial economy, to today's service economy.[2]

In Pine and Gilmore's experience economy, a product's value falls somewhere into a continuum of intensity of experience. *Commodity products*—generic and soul-less goods—are the least valuable to consumers. To a consumer, a commodity product is boring but necessary, as ubiquitous and unexceptional as tap water. Dish detergent. A doormat. Dull as they may be, consumers need essential commodity products but likely won't pay a premium for them.

What Pine and Gilmore call "goods," or unique and distinct products, offer consumers more appeal. Goods would be your favorite brand of running shoe or model of high tech bicycle. Above goods are service products. Consumers value services products that provide a tangible outcome such as a tax preparation service or a good haircut.

Pine and Gilmore's continuum explains the strong appeal of products and technologies that enable DIY innovation. In their hierarchy, the products at the top of the value ladder will be products that offer a consumer an experience, and best of all, transformation. An experience product provides a feeling, like a movie night or a spa treatment. Transformation products change the customer in a deep way, producing a benefit with a positive long-term impact: for example a college degree, a few months at summer camp, or the acquisition of a new skill.

In an experience economy, successful companies with the best profit margins will be those that sell their customers products or goods that offer an experience or transformation. Both experience and the feeling of transformation are compelling and memorable. Consumers pay a premium for such things and will come back again and again.

DIY innovation offers a rich set of experiences, of transformation, providing its practitioners with a sense of community, the opportunity to acquire new hard-won skills, the satisfaction of designing and manufacturing something. 3D printing technologies enable us to transcend the mundane, to break out of the realm of commodity products and dull experiences. Several of today's leading companies that sell 3D printers or related services tap into a consumer need for intense, memorable, and transformative experiences.

## MakerBot

MakerBot Industries offers its customers an alluring blend of creativity, belonging and technical challenges. MakerBot sells consumer-level 3D printers with catchy names like "Cupcake" and their more current model, "The Replicator 1 and Replicator 2." Bre Pettis, one of the company's founders, is a marketing mastermind who has skillfully shaped the company's image and appeal.

A friend of mine relishes MakerBot's ability to transform a relatively dry additive manufacturing machine into a sociable, creative playful endeavor. "When you think about MakerBot, you just feel like for some reason, you want to *hug* them," is how he explained it. MakerBot's loyal and passionate customers mingle on MakerBot's website, `thingiverse.com`, where discussions reflect the company's playful DIY culture.

Like lifestyle magazines that offer readers recipes geared for specific holidays, MakerBot's blog offers printing projects and tips to help its readers celebrate Memorial Day, Christmas, and the Fourth of July. MakerBot observes special holidays, for example, "Geek Pride Day." To celebrate Geek Pride Day properly, users can download design files to print their own special towel hooks, a figure of Han Solo trapped in Carbonite, a pair of black hipster glasses, or Geek key chain.

**New 3D printers being prepared for shipping in MakerBot's Brooklyn-based BotCave, 2012**

At the time of this writing, MakerBot was wrapping up a period of intensive, frenzied growth. It was moving its headquarters out of the grungy streets of Brooklyn's industrial area into downtown Manhattan. It's easy to mistake MakerBot for an overnight success thanks to their sudden media attention. Yet its founders, Adam Mayer, Zach Smith, and Bre Pettis toiled unheralded for a few years, supported by money from family and friends.

When I visited MakerBot's headquarters, the scene was one of high energy chaos and runaway growth. MakerBot's founders have done a masterful job of transitioning the company from a three-person startup with deep DIY roots, to today's media-savvy and well-funded company. Thanks to MakerBot's playfully appealing public company persona and a game-changing appearance on the Stephen Colbert show (where Adam, Bre, and Zach made history by 3D printing a plastic replica of host Stephen Colbert's head), MakerBot is enjoying ever-increasing sales revenue.

MakerBot's genius lies not in its technology. To keep prices in a consumer-friendly range, its Cupcake and Replicator 3D printers employ technically primitive printing mechanisms and print in plastic. MakerBot's genius lies in its spirit of playfulness, the company's ability to make design and manufacturing a unifying, fun, and transformative experience, for its customers.

## Shapeways

As powerful tools of design and production become available to everyone, the lines between professional and amateur, buyer and seller, designer and consumer begin to blur. If MakerBot is the purveyor of 3D printers and online communities that embody free-spirited creativity, Shapeways represents an innovative design marketplace. Shapeways is a web-based community/marketplace that hosts storefronts for designers and 3D prints things for customers who send in a design file.

Shapeways was based in the Netherlands, but in 2011 moved its headquarters to New York City. Like MakerBot, Shapeways is growing rapidly. At the end of 2012, to the delight of local politicians who cut the ceremonial ribbon with a 3D printed pair of scissors, Shapeways opened a factory of fifty 3D printers in Long Island City.

We visited the Shapeways's New York office, which like MakerBot, was a scene of cheerful and rapid growth. The lobby of the tall mid-Manhattan office building was littered with wooden scaffolding and someone had removed company names from the lobby and elevator. After a few fruitless journeys up and down, we found Shapeways on the ninth floor.

Robert Schouwenburg, co-founder of Shapeways, met us at the door. I had met Robert in person a few years ago, in Eindhoven, when Shapeways was just a few employees strong and its "factory" consisted of half a dozen printers in a building in a rural industrial park. Like a good host, Robert greeted us and offered us a hot cup of espresso (served in a 3D printed cup that boasted six small handles).

**A custom espresso cup printed (and in use) at Shapeways**

Before Robert co-founded Shapeways in 2007, he worked as a "white hat" hacker at a major consulting firm, breaking and entering company computer networks to identify security holes. The leap from computer networks to running a 3D printing services firm makes sense if you see Shapeways from Robert's perspective, as a "platform for personal fabrication that works for the consumer and for the professional." Robert sees the future of consumer 3D printing as one that's based on the notion of a service-oriented "platform" rather than at-home fabrication.

"Shapeways is like Amazon's platform, an outlet for people to sell their own products," Robert explained. "In 2002, Jeff Bezos sent Amazon employees an internal memo that Amazon was going to be a platform company." Bezos's directive that would change the future of ecommerce was that every team inside the company must expose their data to one another and to the outside world via APIs. Over a decade later, Amazon is the Internet's biggest e-commerce platform with automated feeds to outside product vendors, inventories, and other software companies.

While platform strategy may steer Shapeways's business model, its magic lies in the artistry of its designers and the broad range of materials that people can print things in. If you buy your own 3D printer, your choice is between different colors of plastic. At the time of this writing, Shapeways offered its customers the choice of 25 different printing materials. Shapeways sells everything, from custom charms for a charm bracelet from designer Mark Bloomfield of ElectroBloom to the abstract artistry of Francesco De Comité. Its online shop owners are starting to earn real money, the combined equivalent of a few hundred thousand U.S. dollars a year.

So far Shapeways designers and buyers haven't yet ventured into the stuff of everyday manufacturing, for example, printing car parts or making injection molds. Somewhat closer to the edge of real manufacturing lies another pioneering DIY business called 100kGarages that spans the worlds of consumers, skilled amateurs, and moonlighting professionals. 100kGarages's founders Ted Hall and Bill Young have built an online community where makers and buyers can have custom products made "just the way you want them."

## 100kGarages

"Nobody's making a full-time living yet by making things for people," said Ted Hall. Many active users are professional small manufacturing firms that own a CNC router or 3D printer and want to make some extra money. "However, what we hear—from both makers and buyers—is that they find one another on 100kGarages, build up a relationship, and then get together for bigger contracts offline."

100kGarages began as a partnership between Ponoko and ShopBot Tools. Consumers post projects on the site. Makers and professional manufacturer bid to make them. Ponoko is a New Zealand-based aggregator of custom designs and fabrication services that likes to describe itself as "The world's easiest making system." ShopBot Tools is based in the United States and makes and sells CNC routing manufacturing machines.

Although the Internet has made it possible for specialized communities of people to connect with one another, Ted and Bill are finding is that people still gravitate to people nearby, if they can. 100KGarages customers do not choose their maker according to their physical proximity. But, Ted explained, "if someone has designed something, they like to use a local maker to make the prototype. And if they end up selling their product, they look for shops near their customers so they can use local materials."

Easy payment technologies like PayPal lubricate the online transactions between site users. Small shops can not afford to buy materials upfront. "Most of the makers prefer to be paid before they invest in materials and time," explained Ted.

Another enabling technology that aids the Internet-based economy is online user ratings. "We don't have a certification process," said Ted. "The best way for a maker to certify himself or herself is to post a gallery of their work on the site. That plus online user ratings are a pretty effective system."

## A future economy of printable products

New technologies and business models sweep across the Internet as quickly as a forest fire sweeps over a rain-starved national park. In contrast, innovation in manufacturing is is a slow and cautious process. Bruce Kramer is a National Science Foundation program manager and a long-time evangelist of 3D printing and personal-scale manufacturing tools. His experience is that in manufacturing, innovation is a high stakes game. Too high. "To innovate, manufacturing needs new technologies that enable it to become less risk-averse and more like the internet and software communities in outlook," he explained.

Today's mass manufacturing industry is the backbone of the world's economy. Its sheer scale, complexity and physical logistics make innovation risky. Big manufacturing companies must experiment carefully. The core ethos of manufacturing is slimming overhead costs, keeping within the boundaries of environmental and workplace regulations, and efficiently moving physical goods from one place to another.

3D printing lowers the risk and cost of introducing novel products to the marketplace. Less investment upfront enables small manufacturers to make a few products at a time in response to customer demand, and scale up production of only those products that sell. For example, producing a small cell phone cover using traditional injection molding requires investment in a mold that costs $10,000 at minimum. This initial investment can be amortized only by a relatively large number of sales. More importantly, this overhead creates an "innovation friction" that prevents small ideas from ever being tested until the overhead has been recovered. But as any evolutionary biologist will tell you, big breakthrough innovations are often composed of a succession of lots of small ideas.

With 3D printing, trying out lots of small ideas becomes possible. A small company or individuals assume less financial risk if they can make and sell an untested product in small volumes to see how the market responds.By starting small using 3D printed production, a new venture does not have to invest in the machinery and infrastructure associated with today's manufacturing environments.

## Scale up from one: don't quit your day job

Innovation friction is not just a matter of concern for large companies; it is especially pertinent to individual inventors. New business models in any industry are activated by the democratizing of tools of production. Personal manufacturing tools liberate entrepreneurs from intensive capital resources and skill. Future entrepreneurs will be able to experiment with more new products and more new business models than ever before with almost no upfront financial investment.

Image courtesy of Mark Kendrick

**Mark Kendrick designs and sells unique model train parts in stainless steel. This is a 3D printed "cow-catcher" sold on Shapeways.**

Entrepreneurs with ideas for new businesses will be able to prove their idea to themselves and to potential investors, or maybe even postpone the need for investment altogether. This business model, often referred to as "scale up from one," democratizes manufacturing and retail by lowering the barriers to entry.

Launching a new product today requires its maker to start large: industrial-scale manufacturing machines are not designed to make only one item at a time. Would-be entrepreneurs must procure large amounts of material, invest in factory machine time to tap into economies of scale, and pay for the necessary retail shelf space.

With 3D printing, people can keep their day jobs while they explore the market potential for a novel product they dreamed up. Small companies are spared the financial risk of investing in costly machinery, shelf space they may never use, and expert technical help. People living in subsistence economies in developing nations without access to capital will be able to start manufacturing at local fabrication centers without needing of capital to pay for infrastructure that may never be used.

3D printing lowers the cost of entry into the business of making things and will offer entrepreneurs a cheaper, less risky route to the market. Instead of raising money to set up a factory, entrepreneurs could run off one or two samples with a 3D printer to see if their idea works. If the first samples sell, they could make a few more, even make changes to the design if buyers ask for them. If demand continued to grow, they could scale up production into a traditional factory or invest in a larger number of 3D print runs.

## Pay-per-print: FabApps

Future designs for 3D printing will spring from unlikely places. What if you want to design a and 3D print better hairbrush? It's unlikely that you could design a good, ergonomic, and safe brush—even though a brush seems like a simple product, it takes years of experience and know-how to design a successful one. Yet, with the advent of 3D printers, people who have no prior experience will design their own product.

One future source of ready-designs for printing may be downloadable fabrication applications, or FabApps for short (a term coined by my former students Daniel Cohen and Jeff Lipton). A FabApp, like an iPhone App, would cover a narrow range of applications, yet would offer its buyer just the right balance of customization and ease of use.

A FabApp would cost 99 cents. You would buy a FabApp for a particular need, for example, to create a set of custom grips for your bicycle handles. After you purchased a FabApp online, it would guide you through the design process. You would upload a photo of your hand and a quick optical scan of your existing bicycle handles (to ensure the perfect fit). The last step: choose a color and material, click "print" and in a few minutes you would become the proud owner of a brand new set of perfect-fit custom handlebar grips.

FabApps will offer expert designers of the future a new business model for integrating their design expertise into a growing economy of distributed manufacturing. Like iPhone apps, FabApps will generate a new economy. Small custom printing apps will find their niche in narrow yet complex markets that are too small to attract the attention of big manufacturing companies, yet large enough to offer opportunity to small businesses and individuals.

## Continuous customization and product variety

Why do some technologies shake up our world while others quietly join us without affecting our daily lives very much? Bursts of innovation happen when an emerging technology removes a once prohibitive barrier of cost, distance, or time. 3D printing shrinks two prohibitive costs to zero: the cost of customization and the cost of complicated shapes.

Since customization is costly, product variety is costly. Companies can't afford to offer their customers all the variety their customer would like. Figuring out what sort of product variations customers prefer is a costly and error-prone process.

Focus groups are slow and expensive. Benchmarking competitors might work, but it's still uncertain. For complex products that offer lots of options and compete in constantly shifting markets, traditional modes of market research are outdated by the time the analysis is finished.

Startup companies, in particular, can't afford traditional market research tools. In his book *The Lean Startup*, Eric Ries suggests that companies should explore and experiment with multiple new ideas at the same time and adjust their strategies on the fly.[3] Ries argues that startups should conduct a continuous steady stream of small, lean experiments.

3D printing will help companies quickly market test new products and adapt to market feedback. By 3D printing customized versions of a product, a business can circulate several options amongst its customers. Traditional factory produced product variation is expensive. Even what would be considered small batches in a factory would be too large for this sort of iterative and continuous market testing.

Imagine that you just founded a new startup. You're selling a software product and you're not sure how much customers are willing to pay for your product or what features they want. Ries suggests that a startup offer its product in different flavors at different price points to different customers. Then the startup should gather data, shift variables, and try again. Patterns in collected data should reveal a product's best features and prices.

As always seems to be the case, with digital products, this sort of dynamic experimentation is easier and cheaper. Data on user preferences and purchases is more readily available. For physical products, an iterative, real-time approach to testing product variables is harder to implement and user data more difficult to gather.

How could Lean Startup principles be applied to a physical product? Imagine your startup sold cell phone covers. You wanted to compare one cell phone cover that had a star shaped pattern embossed into the back, and another, plainer version that didn't. You could display both for sale and see which one was purchased more but you'd actually have to manufacture both and pay for the cost of producing two different injection molds.

With a 3D printer, you could produce a thousand different covers as easily as a thousand identical covers. You could offer a large variety of covers on your website, and see what sells.

You might never sell some of the covers. You might sell only a few of some of the others. Eventually you would learn which particular cover turned out to be hit. That's the cover that would justify the investment.

A month later, you could repeat this experiment, eliminating those that never sold and creating newer variations of those that did. Similar to Darwinian evolution and natural selection, you could evolve your physical product in a continuous process.

Such a process is only possible with the economy of scale offered by a 3D printer—driven by low lead time and upfront investment. It's a new way to quickly find out what works and what doesn't, what sells and what doesn't, and what improvements are needed and which aren't. It's all about quick adaptation.

Variety can be a double-edged sword that needs to be used with caution. As 3D printers open the door to offering a larger variety of physical products, we may begin to experience a physical variety overload. Offering too many options to customers can create an overload of choice.

## Sex and entertainment

Small-batch production of custom products sold at premium prices, is ideal for: the sex industry. Sex and porn fueled the growth of the early Internet. In fact, some people argue that one of the biggest motivators behind the rapid improvement of streaming video technology (and the VCR before that) was to offer porn customers a better experience. Video gaming is another economic giant. Some argue that gaming technologies are the key driver behind improvements in graphics display technologies.

3D printing's effects may someday be felt in the sex and video gaming industries. People don't want to admit it in front an audience, but frequently after I speak, somebody tells a story about finding a rogue sex toy left behind on their company's 3D printer. Or they describe the time they discovered the reason their printer kept running out of a certain polymer printing material was because a staff member was printing naughty figurines after hours, when everyone had gone home.

Wherever there is a deep human need, there is money to be made. It's a short step to a business model. But leading companies in the 3D printing space have, so far, deliberately stayed out of the sex market.

It's going to be interesting to watch when the first 3D printing sex shops emerge. People's creativity will know no bounds if offered the ability to fabricate unique novelty toys and related gear in the privacy of their own homes. Combine that with scan technologies, easy design software, and robotics, and the possibilities are endless.

## Microfinancing microfactories

Microloans, microcredit, and microtransactions are part of a burgeoning microfinance economy. Micro scale transactions are possible thanks to communications technologies and online banking that quickly handle international transactions. Lots of small investments can be easier to gather than a few large investments. In certain regions, several small investments can have a larger cumulative impact than a few carefully chosen larger investments.

Traditional lenders often prefer to invest in borrowers with an established track record of performance and credit history. Yet some of the most gifted entrepreneurs, inventors and makers often lack an established track record. Microloans may have enormous potential to alleviate poverty.[4]

In many communities worldwide, certain segments of the population such as women, minorities, and the poor are less likely to receive loans or investments for innovation. More readily available microloans have changed this. As of 2009, an estimated 74 million people have received microloans totaling $38 billion. The reported repayment rate was 95–98 percent, which is higher than the repayment rate of some established banks or even countries.

The microfinancing strategy is directly comparable to a future business paradigm which we call microfactories. The name microfactories describes their small physical footprint and small production volume. However, microfactories also run on microinvestments, reduced time commitment, and the all-around decentralized nature of their operations. Microfactories could have the same positive economic impact as did microloans, transforming low-income communities and empowering disenfranchised sectors of the population.

Related social trends will enable microfactories to grow. Open source collaborative business models will allow microfactories to access necessary expertise and tools. Online crowdsourcing will allow microfactories to explore new ideas. Crowd financing will allow microfactories to raise funds to finance new ideas.

In the future, economically advanced nations will no longer be able to base most of their economies on profits earned from mass manufacturing. Low wage manufacturing jobs and the production of low margin commodity products will continue to migrate to nations where labor costs less. 3D printing technologies will enable companies to build new business models, to carve out a profitable niche in the economy of the future. To survive, companies must climb up the value chain by making and selling high margin products and services that offer consumers a personalized, transformative experience.

# 5 Printing in layers

This chapter is for the technically inclined, for those who want to dig into the mechanical mysteries of the 3D printing process. Otherwise, the brief explanation of printing technology we've already provided is all you need to enjoy the rest of this book so feel free to skip ahead.

The formal industry name for 3D printing, additive manufacturing, is actually quite descriptive of how these machines work. "Additive" refers to the fact that 3D printing methods fabricate objects by either depositing or binding raw material into layers to form a solid, three-dimensional object. "Manufacturing" refers to the fact that 3D printers create these layers according to some kind of predictable, repeatable, and systematic process.

A 3D printer can be small enough to fit into a tote bag or the size of a small mini-van. Printers can range in cost from a few hundred dollars to half a million dollars. Their unifying trait is that they follow instructions from a computer to place raw materials into layers to form a three-dimensional object.

## A manufacturing process at heart

At its heart, 3D printing is a manufacturing, not a printing, process. That's why we were intrigued when we heard that a global company, ABC Imaging added 3D printing to its service offerings. To learn more, we contacted John T. Lee, the man who manages ABC's 3D modeling and rapid prototyping services. John agreed to walk me through the 3D printing process at ABC's headquarters in Washington, DC.

"ABC got into 3D printing because our clients asked for it," John told me. For years the company created paper blueprints and other printed products.

Now clients prefer three dimensions. "We make architectural models and product prototypes for architectural and engineering firms. Our clients like to have something physical they can show to a client and pass around the table."

ABC Imaging has been in business since 1982. About 5 years ago, the company hired John to run their growing 3D modeling and rapid prototyping business. John studied geology and geophysics at Rice University in the 1990s and was introduced to 3D printing early in his career when he worked at a firm that produced three-dimensional geographic maps.

ABC Imaging's DC-based headquarters gleams with pride and attention to detail, from its carefully organized website to the immaculate glass-topped Scandinavian-style table in its main conference room. From a single print shop in DC, the company has grown into 35 hubs and 550 employees in cities worldwide. After we completed our introductions, John led me back into ABC's production area. ABC's main printer room was as big as a good-sized classroom with a comfortable neo-industrial flavor. ABC employees in business-casual clothing tended several whirring large-bed industrial-scale 2D color printers.

I asked John whether he thought of 3D printing as a printing process or more like manufacturing. John paused and said, "I think that the name '3D printing' is almost a marketing term. 3D printing is manufacturing—it can be a dirty and messy physical process. We use chemicals, and depending on what I'm doing in here, sometimes I wear a gas mask."

ABC Imaging owns several different models of industrial-grade 3D printers that are scattered around the world in different company locations. In the DC headquarters, the company's small 3D printing "factory" was tucked into the corner of the main production area behind several staff desks and cubicles. This brightly lit, glass-walled room was once the company's kitchenette and staff break room. Two medium-scale industrial 3D printers covered most of one wall.

John runs ABC's printing services using high-end printers that fabricate a detailed, color model in a day or two. Production timelines vary, depending on the complexity of the model and whether the customer's design file is a rough, untested first draft or a watertight and, hence, printable design. Most printed prototypes are made in white. Some models and maps, however, are printed in color.

In ABC's printing room, a boxy machine sat next to several 3D printers. This boxy machine had a glass front and two arm holes on each side. Inside, behind the glass panel on the front, rested a powerful air gun. John showed me how to put both arms into the holes on the sides of the box and blast loose powder off a freshly printed object using the air gun. On the opposite wall, several small, metal tubs bubbled with solvents where printed objects were dunked to finish their surfaces and dissolve away any residual powder.

John showed me a glowing white architectural model of a mansion fit for an emperor. The mansion's front boasted eight pillars that framed a grand veranda. On its roof, delicate handrails encircled a flat rooftop deck presided over by a dome with finely etched radial lines. On the mansion's right side, a set of curved steps carried visitors to the front door.

Such an elaborate architectural model could not have been made using traditional plastic manufacturing techniques such as injection molding. Nor could a carving tool (or person) have the finesse to cut its delicate shape from a block of solid plastic. Traditional cardboard models would not do much good either. The mansion's rooftop deck rail is as finely detailed as a spider's web. The open space behind the pillars of the front veranda, if carved, would result in the pillars snapping under the pressure.

John was blasé about his printed creation, having fabricated far more elaborate objects. "3D printing lets you make some amazing models and parts. Some of the geometrics you can't make by any other method," he said.

Image courtesy of Midwest Studios. Photo: Ed Watson

**The ultimate 3D printed architectural model: an ancient monastery**

## Two families of printers

I generally explain to people that there are two major families of 3D printing technologies. The first family of printers deposits layers of raw material to make things. The second family of printers binds raw materials to make things.

The first family—let's call them "selective deposition printers"—deposits raw material into layers. This class of printers squirts, sprays, or squeezes liquid, paste, or powdered raw material through some kind syringe or nozzle. 3D printers used in people's homes and offices are usually of the deposition type because lasers or industrial-grade heat guns can be too fragile and dangerous.

The second family of printers that binds (does not lay down or deposit) raw material typically trains a laser or adhesive onto some sort of raw material. This class of printers—called "selective binding printers"—use heat or light to solidify powder or a light sensitive photopolymer. If you remember the bold claims of the Cubital salesman, he told my classmates and me that he had made his demo on a machine that "printed" it out using a laser.

### Printers that squirt, squeeze, or spray

Let's first explore selective deposition printers that deposit some kind of raw material through a print head or nozzle. The raw material for printing might be soft plastic that will harden once it hits the print bed, raw cookie dough, or even living cells in special medical gel. If you've seen a consumer-style 3D printer in the media, such as MakerBot, you've probably seen this type of printer.

The formal, technical name for the printing technique used by this category of printer is "fused deposition modeling," or FDM. FDM printers were invented in the 1980s by Scott Crump who then built a company on the technology. If you see a machine described as an "FDM printer," that means it squeezes out some kind of soft raw material through a print head.

This type of 3D printing process begins several steps before the print head kicks into action. The first step is to find a software design file that will tell the 3D printer's built-in software (also known as "firmware") what it needs to print. Once the design file is ready, users attach their laptop to the printer and save the design file into a special file format that the 3D printer's built-in firmware can read and work with (we'll explain the details of this conversion process later).

Photo courtesy of Stratasys Inc.

**The white object is being printed. If you look closely you can see the red conical tip touching the top of the white part and extruding a thin strand of white plastic. The dark plastic is printed support structure that will wash away later.**

Once the printer's firmware reads the formatted file, it calculates the mechanical path and actions of the physical print head. For example, the print head needs to know where its print nozzle should deposit the outline of the design's shape, and how much material the print head should squirt, where, and so on.

Once the 3D printer's firmware finishes planning the sequence of operations, the physical printing process can begin. Printers that deposit material typically move the print head along a set of horizontal and vertical rails (what engineer's call a "gantry") that zips the print head where it needs to go. To deposit the first layer, the print head outlines the shape of the footprint of the object being printed. This first layer, like a pencil tracing around the bottom of a coffee cup, outlines the base of the object. The print head will then proceed to scan back and forth to fill in the contour, like a child filling in a shape in a coloring book.

After the first footprint layer is printed, the print head is raised slightly and returns to work to lay down the second layer. The printer continues to repeat this process, patiently laying down one cross-section of the object after another, as depicted in the design file, a process that can go on for hours, even days.

The good thing about this branch of the 3D printing family tree is that their printing technology can be simplified into relatively low-tech versions that are low-cost and can use a wide range of materials. Any raw material that can be squeezed through a nozzle can be 3D printed. Frosting, cheese, and cookie dough are a popular raw material for food-loving print enthusiasts. Another emerging printing material is "living ink." Living ink is a blend of living cells placed in a special medical gel that medical researchers use for bioprinting research.

Although manufacturing and design companies use large and expensive printers of this sort, selective deposition printers are ideal for home, school, or office use. Even low-end printers of this sort operate quietly, and the fact they use a relatively low-temperature print-head makes them safer to operate than printers that use high-powered lasers.

A major downside of selective deposition printers is that they can print only in materials that can be extruded or squeezed through a print head. Molten metal or glass, for example, must be shaped under different conditions. Most deposition printers on the market today keep things simple by using a special type of plastic that's created especially for them. 3D printing plastic is sold in spooled, spaghetti-shaped strands whose end is fed directly into the printer where the plastic is melted and squeezed out through the print head.

### *PolyJet printing*

PolyJet printers are the youngest members of the deposition branch of the printer family, developed in 2000 by an Israeli company called Objet Geometries (which merged with Stratasys in 2012). PolyJet printers borrow technologies from both major branches of the 3D printer family tree, combining a print head that sprays liquid photopolymer into extremely thin layers and firms up the photopolymer with a bright UV lamp.

The benefit of using PolyJet printing is that spraying droplets is a fast and precise way to lay down layers as thin as 16 microns. As a point of reference, the diameter of a red blood cell is about 10 microns. The precision of PolyJet printers makes them ideal for industrial or medical applications, where "high-resolution" shapes and fast printing can be mission critical. PolyJet printers can use several print heads at once so they can print in multiple materials in a single print job.

A major downside of PolyJet printing lies in inherent limitations in the printing material it uses, a type of plastic called a photopolymer. Photopolymers are highly specialized, expensive plastics that respond to UV light. Plastic can be one of the most rugged manufacturing materials there is, but most photopolymers are still relatively fragile and brittle, which limits their range of applications.

**A PolyJet printer fabricating one of the candidate structures for this book's cover page**

### *Laser Engineered Net Shaping (LENS)*

Another member of the selective deposition printer family, Laser Engineered Net Shaping (LENS), blows powdered material into a carefully guided high power laser beam. Some of the powder misses the beam and falls aside, but the lucky particles that hit the focal point of the laser get instantly melted and fused to the growing part surface. Thus, as the laser focal point scans the contours of the object and the nozzle blows more powder, the part gradually grows, layer by layer.

The advantage of this process is that it can make objects from hard materials such as titanium and stainless steel. Until such metal "printing" processes were invented, 3D printing wasn't taken that seriously by "big" industries because it could only work in plastics (polymers). When metal printing processes like LENS emerged, big industries such as aerospace and automotive became more interested in 3D printing. LENS technology is used today to make complex objects from hard metals, such as titanium turbine blades with internal cooling channels.

Photo courtesy of Richard Grylls, Optomec, Inc.

**Powdered metal is blown into a laser beam. Particles that hit the focal point melt and gradually build up a metal part.**

Since more than one nozzle can blow powder into the laser beam at the same time, multiple base metals can be used simultaneously to "print" alloys (mixed metals) in tunable ratios. The ratio can even be varied depending on the position of the head, leading to graded alloys.

### *Laminated object manufacturing (LOM)*

Last but not least, another member of the selective deposition printer side of the family are laminated object manufacturing printers (LOM). LOM printers don't use a print head to form layers. Instead, LOM printers, like their name suggests, laminate thin sheets of material into a single solid three-dimensional objects.

The LOM process begins with a design file. Instead of a print head, a knife or laser beam does the work. Following the design file's guidance, the cutting tool slices out the contours of a shape from a thin film of paper, plastic or metal. Imagine taking a coffee cup, setting it down on a sheet of paper, and cutting out its outline in the shape of the cup's base.

After the cutting instrument is done with a sheet, the LOM printer whisks the cut-out aside and lays out a fresh sheet of adhesive film to cut the next layers. The printer stacks together the cut layers of paper, plastic or metal. When the cross sections of the object have all been cut out, the printer laminates and presses the layers of cutouts to fuse them into a solid 3D object. Some models of LOM printers fuse cut sheets of aluminum foil with powerful ultrasound vibrations that cause the sheet to rub against the previous layer and consolidate into densely packed layers.

## Printers that fuse, bind, or glue

The second major family of 3D printers is made up of printers that use a selective binding process to fuse or bind raw material into layers. Many of the earliest commercial printers used this approach. Two variations of this method, in particular, are in widespread use: stereolithography (SL) and laser sintering (LS).

### *Stereolithography (SL)*

Stereolithography (SL) was one of the earliest commercial methods of 3D printing. Imagine a small vat of liquid polymer sitting inside a printer the size of an apartment-sized refrigerator. The printer sweeps a laser beam over the surface of a special type of plastic, a UV-sensitive photopolymer that hardens when exposed to UV light. Each sweep of the laser traces the outline and cross section of the printed shape in consecutive layers.

After each sweep of the laser, a moveable table holding the printed part is lowered a fraction of a millimeter. The part sinks a bit into the liquid, and fresh photopolymer floods its top side. Some SL printers work in the opposite direction by aiming the laser upwards into the photopolymer, then lifting (rather than lowering) the printed object to flood its base (rather than its top) with fresh liquid.

After an object is 3D printed in this method, there's still more work to be done. Excess material needs to be rinsed off and the surfaces sometimes need to be sanded by hand. Depending on what's printed, sometimes further curing is done in an ultraviolet light "oven."

The upside of SL printing is that the laser is fast and precise. Multiple lasers can work in parallel to trace out shapes at a higher resolution than can today's extrusion-style 3D print heads. Today's industrial-scale 3D printers can fabricate precise models and parts in layers as thin as 10 micrometers—thinner than a sheet of thin paper. As the quality and range of raw photopolymers continues to expand, SL printers can fabricate a broader range of objects with specialized material properties.

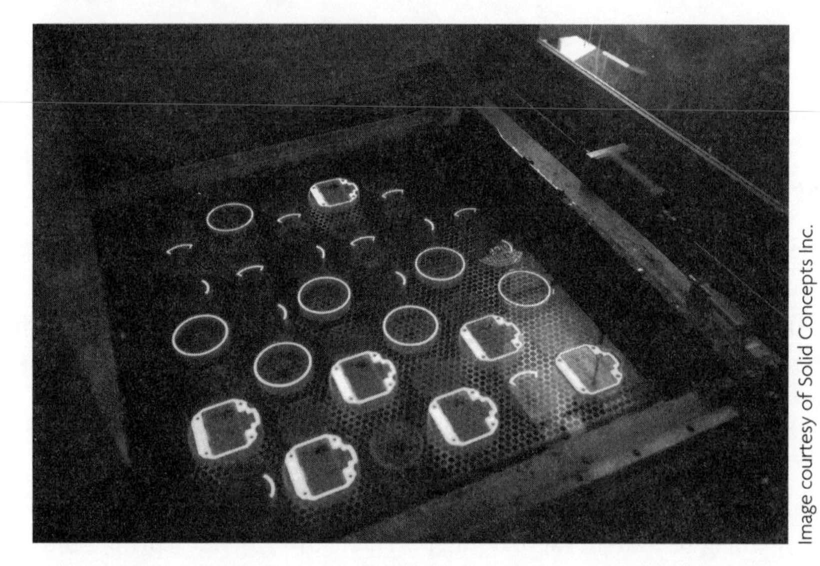

Image courtesy of Solid Concepts Inc.

**As the UV laser traces the shape of successive cross sections, the solidified parts are slowly lowered into the tank.**

A downside of SL printing is that fumes from uncured photopolymers can be toxic to breathe. Also, photopolymers are not as strong and durable as their thermoplastic cousins used in industrial injection molding. The cost and complexity of maintaining a machine that uses lasers makes SL printers too expensive for most home users, but cheaper machines that use low-cost UV lasers from Blu-ray discs may grow this market. SL printers can currently print with only a single material at a time.

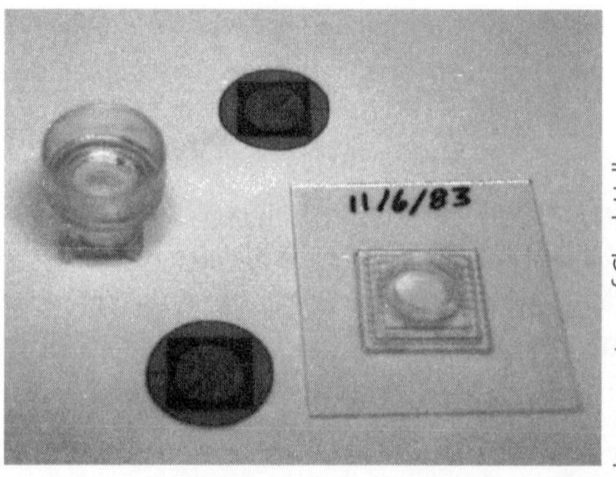


Image courtesy of Chuck Hull

**Here is one of the world's first 3D print jobs. The technique used, SL, dates back to the early 1980s.**

### *Laser Sintering (LS)*

Selective Laser Sintering, usually referred to as LS, was invented in the 1980s by University of Texas researchers Carl Deckard and Joseph Beaman.[1] Selective laser sintering follows a technique similar to SL printers. Instead of using liquid photopolymers in a vat, however, LS printers use powder.

Like SL, the LS printing process is not what many people would envision as a "printing process." These printers trace a high-power laser beam over the surface of a bed of powder. The powder melts where illuminated by the laser. A roller inside the printer brushes a fresh layer of powder on top and lowers the print bed a fraction of a millimeter.

Printing with powder instead of liquid materials has its advantages. An object printed in powder is less likely to collapse during the printing process since the unfused powder acts as a built-in support. In some cases, unused leftover loose powder can be recycled and used in another print job. Powdered material is more versatile because many raw materials can be obtained in powdered form, including powdered nylon, steel, bronze, and titanium.

On the downside, LS printers create objects whose surface tends to be porous rather than smooth. LS printers currently can't print different types of powders at once. LS printers aren't yet a good fit for home or office use. Since some powders have a tendency to explode if handled incorrectly, an LS printer requires a sealed chamber filled with nitrogen.

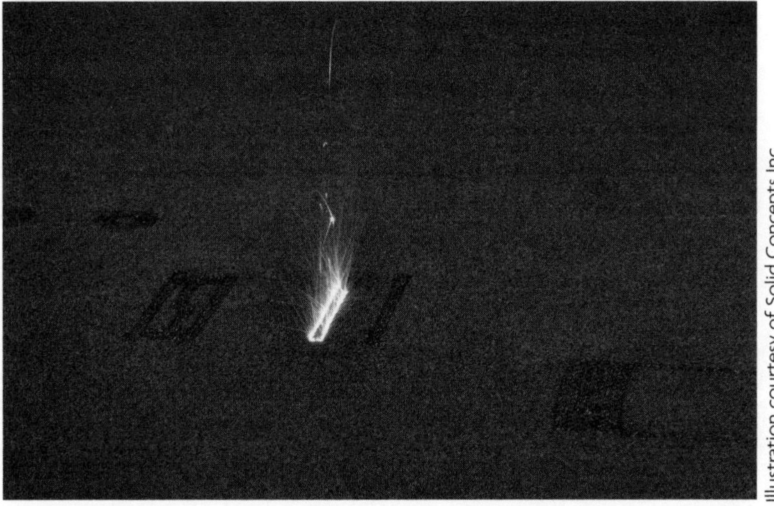

Illustration courtesy of Solid Concepts Inc.

**A laser beam melting and fusing powdered metal. The printed object ends up buried under powder when the printing process is complete.**

Finally, the LS printing process is hot. A completed printed object can't be grabbed quickly out of the machine. Depending on the size and thickness of the layers, larger parts might need to cool for up to a day.

### *Three Dimensional Printing (3DP)*

3DP, in yet another confusing naming convention, use a process called "three dimensional printing" where the print head squeezes adhesive—or some kind of glue—onto raw powdered material. 3DP was invented in the late 1980s by an MIT student named Paul Williams and his advisor Professor Eli Sachs.

At that time, commercial additive manufacturing systems used lasers and sometimes toxic printing materials and were the size of a small truck. Early additive manufacturing machines were complicated to operate and expensive. Since 3DP was a welcome alternative, MIT would later patent the technology underlying 3DP and license it to several companies (where it became the foundation of many of the world's commercial 3D printers).

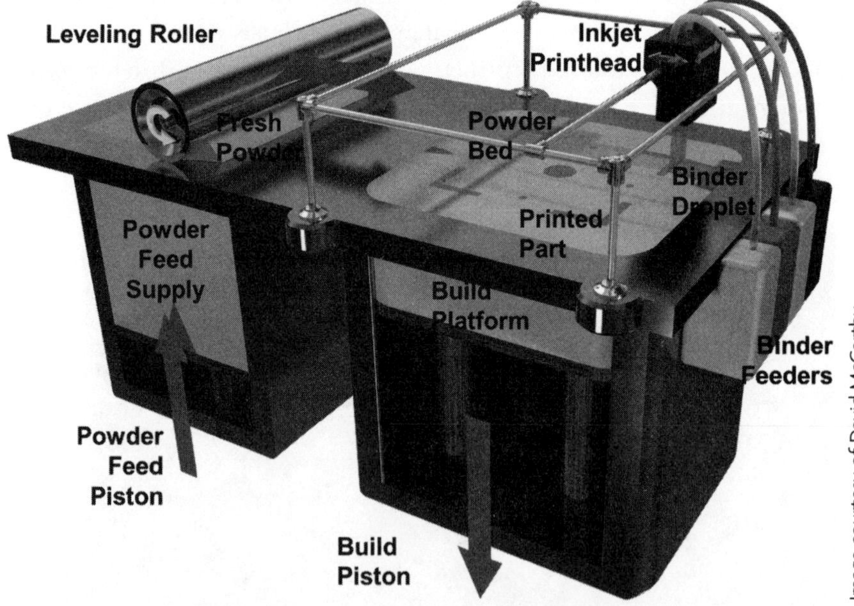


Image courtesy of David McCarthy

**The color 3DP process inkjets colored glue onto a starch-based powder bed, then spreads a new layer of powder and repeats.**

The breakthrough of 3DP printing was its simplicity. Paul Williams' vision for 3DP was bold, particularly given the state of the art at that time. Paul wrote in his master's thesis, "The goal of desktop manufacturing is to fabricate parts at the press of a button, with no further action required."[2] He envisioned 3DP as a desktop manufacturing system that would be precise, fast, cheap, and easy to use.

Today, 3DP has lived up to its creator's vision and has become a popular, low-cost method of 3D printing. Since 3DP printers form layers by squeezing glue onto raw material to make layers, these machines do not use lasers and can work with a broad range of raw materials. 3DP printers don't have to support high-powered components so they are energy efficient to operate. On the downside, because it's difficult to create extremely thin layers without a laser, objects created on a 3DP printer tend to have a rough surface.

One of the biggest advantages of 3DP is its ability to print in color. When glue is deposited, a few additional droplets of colored ink can be ink-jetted as well, allowing the fabrication of full color 3D models. 3DP processes can also be used with a variety of powdered materials, ranging from starch-like material that results in a sandstone-like textured object to powdered clay that needs to be fired in an oven to harden. Some have used 3DP with glass powder, ground bones, shredded tires and even sawdust. Some printers use powdered metal, such as bronze. The glued bronze then needs to be sintered in a furnace to become solid.

## Cleaning up design files

The printing process begins with a design file. Like the bulk of an iceberg that lurks underwater, a substantial amount of preparation involves preparing the design file and setting up the printer. During my visit to ABC Imaging John told me that one of the biggest (but frequently unrecognized challenges) is helping his clients format their design files properly.

A design file must be able to properly converse with the software that's built into a 3D printer. The printer's built-in software (or firmware) tells the printer's mechanical components what to do. Preparing a completed design file to be 3D printed is not always a straightforward process.

At ABC Imaging, the 3D printing process begins when the company's 3D clients—mostly architects and engineers—give John a design file. Most professions tend to favor a particular type of design software. Many architects design in Google Sketchup, for example. Sketchup is a no-cost, easy-to-use design tool popular with educators, architects, and beginning designers. Engineers tend to create design files using high-end commercial solid modeling software. John frequently works with geographers, surgeons, or map-makers whose data originates from an optical scanner, a remote sensor or a medical image.

A recurring challenge in the 3D printing process is compensating for the fact that most design software was created without 3D printing in mind. Design files come in a bewildering array of different file formats, each with their own quirks and challenges. During my visit to ABC Imaging, John told me, "In the 2D print world, if you have a document that's poorly done or unattractive, if you click 'print,' you still end up with some kind of printed document, even if it's not what you really want."

John continued, "With 3D printing, it's not just garbage in and garbage out. It's garbage in and nothing comes out. In the 3D printing world, if you get a design file that's poorly done, you end up with nothing. Or, worse than nothing, you end up wasting expensive raw material."

Most design files, particularly those for complicated objects, require some expert tweaking. "Although people talk a lot about the quality of design software, what really matters is the skill of the human who made the design file," said John. "A messy design file can slow down the process since I need to re-do it and fix it for them."

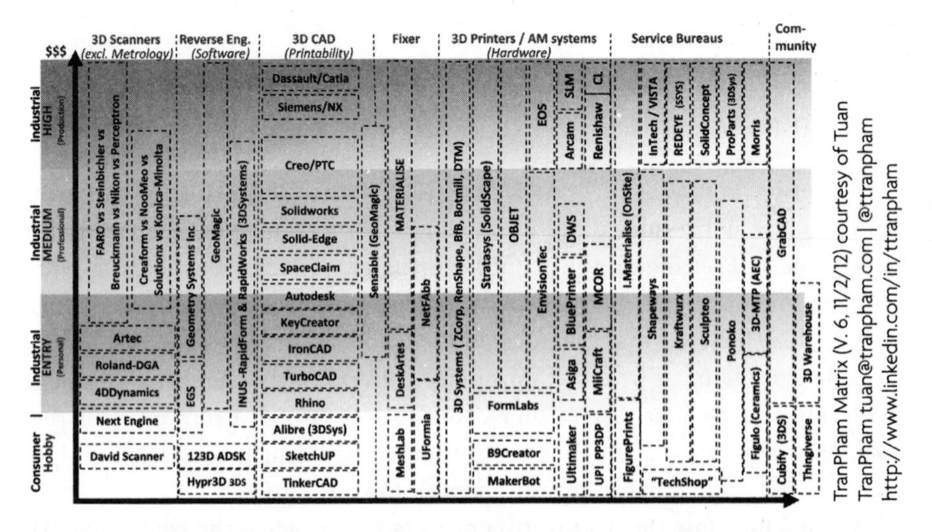


**The 3D printing landscape is a complicated ecosystem made up of products (and open source) tools and applications that span software, hardware, and services and range from consumer to industrial customers.**

After clients submit their design file John's next step is to convert the file into a special format for 3D printing called STL. STL is an industry-standard, decades-old file format that was originally created by Chuck Hull, the inventor of SL printing and founder of 3D Systems. The STL file format is somewhat analogous to the PostScript file format that translates computer documents into something a 2D paper printer can read and work with.

"A big part of my job is quality control to make sure that the STL file is going to be able to do what it's supposed to," said John. Today's STL file format harkens back to an era when people asked much less of their 3D printers. The job of the STL file isn't an easy one. The STL format has had a long and distinguished life but it simply can't keep up with advances in printing technology and design software.

The challenge lies in the fact that the STL file must somehow translate design complexities and intricate details that may be straightforward in digital form, but can be difficult to convey to a 3D print head. For example, engineering software grew up in the era of manufacturing machines that cut objects rather than made them out of layers. Therefore, reflecting its manufacturing lineage, engineering design software is still learning how to think in additive rather than subtractive operations.

After a design file has been converted to STL, the STL "wraps" the design object's digital "shape" inside a virtual surface, called a mesh, that's made up of thousands (or sometimes millions) of interlocked polygons. Each interlocked polygon (triangles are frequently used) in the surface mesh holds information about an object's shape. Somewhat confusing to non-engineers, in a design file conversion, anything designated as a surface includes any portion of the designed object that touches air. For example, a design object's surface can be on the outside of a designed object, or in its hollow interior.

When the STL conversion is complete, the virtual surface of the newly wrapped STL file must be watertight, somewhat similar to the process of coating a physical object in some kind of waterproofing sealant. A watertight STL file is one whose surface mesh accurately and completely covers and captures the design object's surfaces curves and interior hollows. Like a hole or gap in a sealed and waterproofed pair of suede shoes, gaps in an STL file's surface mesh will cause problems later down the road.

Once the STL file is watertight and ready to go, the bridge connecting CAD and CAM is nearly complete. The object to be 3D printed, now snuggly wrapped in a watertight surface mesh, must be prepared for its final phase: the layered fabrication process. At this point, the STL file does its final bit of work. The STL file is read by the printer's firmware that "slices" up the mesh-wrapped digital design into thin virtual layers that will correspond to the thin physical layers that will shortly be 3D printed.

Each virtual slice of an STL file represents a cross-section of what will become the final, printed object. Remember the traced outline of the base of a coffee cup? That outline would equal to the contour of a single "slice" in an STL file and correspond to a single 3D printed layer. After tracing the contour, the printer will need to raster-scan back and forth to fill-in the interior of the contour, like filling in the shapes in coloring book.

Some 3D printers have a built-in visualization tool that double-checks a CAD to STL file conversion. In the future, intelligent software will make sure a design file can print what its designer intended. In the meantime, experts like John guide all types of design files from their digital incarnation into a sturdy, attractive 3D printed model. "Once a design file is ready," John said, "the hardest work is behind us."

If the humans did everything right so far, at this point, once a design file is ready, the printer should take care of itself. Deep inside the printer, microcontrollers and sensors (similar to print drivers in paper printers) tell the machine what to do to make sure that everything works as it should. John said, "After you click print, the machines can run unattended overnight."

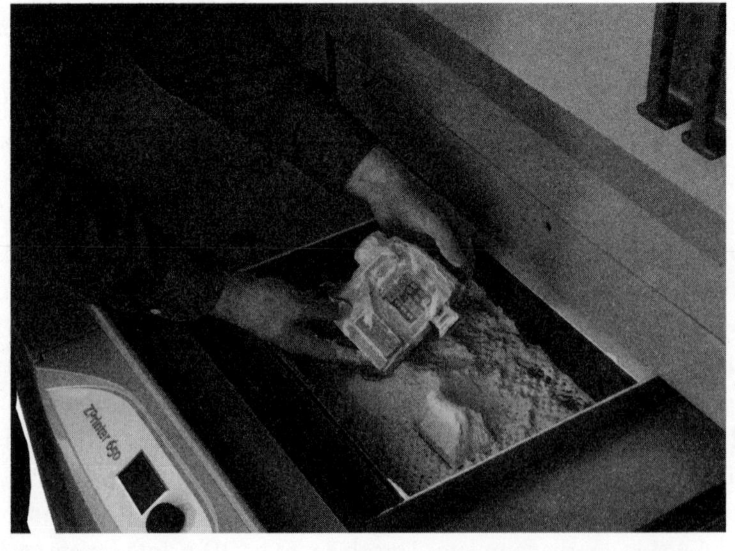

Image courtesy of 3D Systems

**The printed object is unburied from its powder print bed and cleaned up.**

## Printing support structures and post-processing

The printing process at ABC Imaging ends with post processing. When a 3D print job is complete, the humans come back into play. Most printed objects do not emerge from their manufacturing process looking glossy and perfectly formed. After the freshly printed object is pulled from the printer, some human manual prepping and cleanup is required, a procedure called *post-processing.*

Handling complex, freshly printed objects, like other steps in the 3D printing process, has its own learning curve. Newly printed parts can be fragile when they come out of the machine. John said, "Part of the learning curve is that you're going to break a lot of things and burn a lot of material."

The very capabilities that make 3D printing a novel and powerful way to shape new forms and shapes also lend complications to the design and manufacturing process. Objects with lots of spindly parts or boldly looping or hollowed-out designs are the trickiest to print. Imagine 3D printing the Brooklyn Bridge or an architectural model that has a wide but thin, flat roof.

To print "wow" features such as delicate, elaborately interlaced mesh parts, a good designer must consider support structures. Support material is 3D printed alongside the permanent material. Like temporary scaffolding in a construction project, support structures help retain an object's shape during the 3D printing process. Some printers require support structures and some rely on the raw powder to provide support, but some designers will include extra support structures for added resilience. Support materials are removed at the end of the printing process, in the post-processing phase.

The degree and type of post-processing depends the complexity of the object's design, what type of plastic resin it's made from, and how good it needs to look. Removing support material may involve some manual sanding, washing, and smoothing. Some 3D printed objects will be sanded, painted, or welded together with other objects, depending on their purpose.

# The raw materials

If wrestling with design files is one challenge in bridging the digital and physical worlds, the second is forcing raw material to a designer's will. Tiny colored dots of light and rapid streams of binary instructions form the raw material of the digital world. The physical world is not nearly so tidy and easy to manipulate.

Today's 3D printers work mostly in plastic. Although the word plastic has become a synonym for low-cost materials, 3D printing plastic isn't cheap. In fact, the cost of plastic printing material quickly adds up to become a significant part of the cost of running a 3D printer.

Most 3D printer manufacturers provide their own proprietary material. At ABC Imaging when John showed me buckets full of commercial-grade printing powder, he likened the cost of 3D printing plastic to the infamous "razor and blades" business model. "It's like Gillette," he said. "They give away the razor but you can only get the cartridges that fit from Gillette."

The fact that industrial 3D printing technology is optimized to print proprietary vendor-specific 3D printing materials either dampens or drives innovation, depending on who you ask. The downside is that users are discouraged from experimenting with cheaper materials because they risk voiding their manufacturer's warranty. The upside of proprietary materials is that 3D printing manufacturers are eager to invest in developing high-performance and profitable raw materials that will move the technology forward.

Someday print materials will contain living tissue or tiny bits of computing power or will be able to behave in ways that defy understanding. Today, however, most companies and printing hobbyists must content themselves with plastic, metals, ceramics, edible semi-solid foodstuffs, and to a lesser extent concrete or glass.

Plastic is the most commonly used printing material. Plastics engineers divide plastics into two major categories: thermoplastics and thermosetting polymers. The difference between them can be easily remembered by thinking of eggs and cheese. Like cheese, thermoplastics melt when heated. Like cheese, thermoplastics do not change their internal composition when heated, so they can be melted and re-melted many times. Like eggs, thermosetting polymers solidify when heated. Like eggs, they can be used only once; because their internal composition changes when heated, thermosetting polymers can't be melted back down into a reusable liquid form.

Most consumer printers (the kind that deposit raw material through a print head) use a type of thermoplastic called ABS, the same kind used in LEGO bricks. Most stereolithography (SL) 3D printers use light-sensitive thermosetting polymers (the egg-style plastic). Printers that use laser sintering use powdered thermoplastic (the cheese-style plastic).

3D printers can also work with another category of plastics, soft plastics, known as *elastomers*. Like their name suggests, these rubber-band–like materials have various elastic properties. Some of these materials, like silicone, can be squeezed through a syringe and then air-dried. Other soft, rubbery objects can be printed by melting a thermoplastic elastomer, similar to the process used to fabricate hard plastic.

In the early days of 3D printing, skeptics dismissed the new technology as one that had no place among "real" manufacturing machines. Why? Because in those days 3D printers could not manufacture metal parts. Now 3D printers can print steel, titanium, and even tungsten—hard metals that are difficult to shape using conventional manufacturing processes.

Printed metal machine parts are a popular industrial application for 3D metal printing. Several metal printing methods are possible. One method uses a multistep process in which metal powder is first coated with a heat-sensitive plastic binder and then selectively fused together with a laser. Next, unfused powder is shaken off and the remaining metal object is placed into a hot furnace where the plastic glue binder burns off. More direct methods to print metal involve using jets to extrude molten metal or fusing raw metal powder directly with a laser.

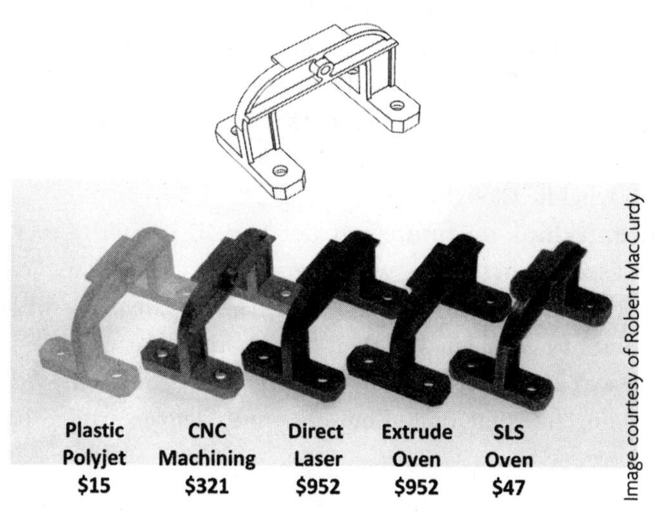


**Design blueprint (top) and fabricated parts (bottom) using a variety of methods and their actual costs**

Home-scale 3D printers can't yet print metals directly. That's changing, however. Consumer-scale printers such as Fab@Home can squeeze out a gel mixed with metal powder. To firm up the printed gel into metal, the printed object is baked in a furnace or kiln. Needless to say, this extra step of kilning is far from simple—it introduces the risk of shrinking, cracking, and warping.

Printed ceramic has the same smooth surface and internal material properties as hand-carved and kilned ceramic. A promising application is 3D printing ceramic bone implants from a patient's CT scans. Ceramic bone implants can be custom-made and, since they're less porous, can be three to five times stronger than those made by conventional methods.[3] Stronger ceramic bone implants reduce the likelihood that micro-debris will break off during surgery, significantly minimizing the risk of post-operation inflammation.

Glass, one of the most common materials used by human civilizations, has been one of the slowest materials to gain traction in 3D printing. Glass is hydrophobic, meaning that it repels water and, therefore, doesn't adhere well. Powdered glass is unpredictable when exposed to heat. University of Washington graduate students Grant Marchelli and Renuka Prabhakar and professors Duane Storti and Mark Ganter have successfully printed objects made of recycled glass in the research lab.[4] Yet, commercial application of glass printing is still mostly for art and jewelry.

Someday, as technologies improve, the 3D design and printing process will be automated and skilled experts will no longer be needed. There's an internal debate in the 3D printing community that's reminiscent of the old "TV in every home" debate over half a century ago. When we were visiting ABC Imaging, I asked John what he thought.

He replied, "I think the future may not be a 3D printer in every home or every office. I see another future, where people and architects and engineers—rather than sending an order to a warehouse to get a spare part—will download a CAD file and have it printed in their neighborhood print shop." He added, "Right now, we're not that far away from that model already. All day long, our bicycle couriers come in and out of here to deliver printed parts to our customers."

# 6 Design software, the digital canvas

Design software shapes our world. Behind almost every architectural model, product prototype, and completed product lies a computer design file. The chair you are sitting on, the stapler on your desk, your car, even the buttons in your shirt were digital before they became physical. Computer design files are the language of modern engineering.

Design software is the heartbeat of 3D printing. Like the pencil and paper hand drawings that guided Victorian shipbuilders through the construction process, a design file tells a 3D printer how to print.

## A word processor for drawing

The first crude, primitive computer-based design tools appeared in the 1950s and were used by researchers and scientists for specialized calculations and crude computer-based simulations. Early commercial design software came onto the market in the 1960s and cost about $500,000 (sold by a company called Control Data Corporation). When I was an undergraduate, we envied the PhDs who "got time" on a CDC mainframe. This gigantic computer took a minute to render a design model, something your cell phone can run thirty times a second.

In 1982, John Walker, the CEO of a small software company called Autodesk wrote an internal memo to his employees. He described his vision for a radical new design software product. He passionately pitched this new product as a low-cost "word processor for drawings" that would run on a microcomputer. At about the same time a few thousand miles away, Chuck Hull was fabricating the world's first crude 3D printed objects.

After much discussion amongst Autodesk employees, Walker named this new design software application "MicroCAD." Today a low-cost, desktop-based design software tool sounds like a viable product. Back then, it was a gamble on an uncertain future. The reason that MicroCAD was a radical design tool was not because it was design software, but because it could run on a desktop-scale computer. Its primary market appeal was its cost, both in terms of the price of the software and the fact that its user didn't have to invest in tens of thousands of dollars in computing power.

Walker envisioned MicroCAD as a design tool that would offer the same performance as its more expensive competitors but for a fraction of the cost. Walker wrote that MicroCAD "[i]nstalled on a desktop computer configuration in the $10K to $15K range, [was] competitive in performance and features to Computervision CAD systems in the $70K range."

In the 1980s three-dimensional modeling was the domain of professional designers and engineers who used computer-based modeling program mainly to stress test machine parts. Microcomputers were raw, new technology, much like today's home 3D printing kits. Microcomputers were puny compared to mainframes or early UNIX servers and lacked the power needed to run complex industrial design projects.

During those heady and innocent days of the personal computing era, the market for desktop computing was just getting off the ground. Today's billionaire titans—who would later make their fortunes selling software for personal computing—weren't yet millionaires in those days and were largely unknown outside of high tech circles. Microsoft and Apple were nowhere near cracking into the Fortune 500. Bill Gates was still flying coach class.

At that time, Walker's fledgling software design company was a loose federation of part-time employees. Each employee was expected to put in at least 14 hours a week in exchange for an annual paycheck of $1 a year plus stock options.

Walker's memo continued "There are no known competitive products on microcomputers today (although there are some very simpleminded screen drawing programs for the Apple, and we must be careful to explain how we differ)."[1] MicroCAD could handle two dimensions and was intended for the creation of architectural floor plans or perhaps to automate the process of drawing blueprints.

Decades later, it turns out that Microsoft, Apple, and Autodesk gambled correctly on microcomputing. As the tidal wave of low-cost desktop computing swept over the landscape in the 1990s, MicroCAD (now renamed AutoCAD) and other commercial computer-aided design software rode along on its momentum. Even a run-of-the-mill cell phone these days has more computing and visual display capacity than a 1970s mainframe computer.

Today Autodesk is a billion-dollar global company. More than 10 million copies of AutoCAD have been sold. The days of "word processing for drawings" are history; modern AutoCAD is a powerful design tool that models in three dimensions.

## Keeping track of x, y, and z coordinates

Design software must seamlessly capture the continuous and geometric essence of the analog physical world and reduce it into discrete binary units. World-renowned physicist Richard Feynman recalled in his memoir a conversation that captured this non-symbolic, non-verbal nature of geometry:

> *One time, we were discussing something—we must have been eleven or twelve at the time—and I said, "But thinking is nothing but talking to yourself."*
>
> *"Oh, yeah?" Bennie said, "Do you know the crazy shape of the crankshaft in a car?"*
>
> *"Yeah, what of it?"*
>
> *"Good. Now tell me: how did you describe it when you were talking to yourself?"*[2]

Feynman makes a good point. The human mind's eye perceives the shape, composition and behavior of the physical world as a collection of an infinite continuum of differently shaped objects. But reducing all of this information into a workable set of symbols that can be re-described to someone else is a challenge.

Design software must reduce our ambiguous and varied physical world into a precise unambiguous "language." Early computers quickly surpassed humans when it came to calculating and keeping track of symbols such as numbers or text. However, it has taken decades and great gains in available computing power to process raw geometry.

Image courtesy of Bathsheba Grossman

**A design file (right) electronically describes the physical object to be printed (left) and guides the printer through its paces**

A computer-based design file captures the shape of physical objects using sets of x, y and z coordinates. When describing the shape of a simple block, it is sufficient to specify its height, width, and depth. But when describing the shape of a more complex shape, such as a flower, say, things can get much more complicated. Computers model the shape of such "general" three-dimensional physical objects as sets of thousands, even millions of trios of x, y, and z coordinates.

When design software shows its user a design object on the computer screen, behind the scenes the computer's processor is working hard calculating mathematical equations that generate the appropriate image. Each time a user clicks on the design in progress or stretches its edges with the mouse, the computer is rapidly running a series of calculations to adjust the object's x, y, and z coordinates.

It's not that hard to keep track of design changes for a simple cube made of a single material. However, things get more complicated when shapes become curved or elaborately formed. Even a simple cylinder that can be described perfectly by its height and diameter alone requires quite a bit of calculation to display correctly in all of its curved, shaded glory. For a design project to create a highly ornamented metal radiator, the design file is going to have to keep track of a massive set of x, y, and z data points to depict each curve, hollow, and edge of its surface.

In the 1990s when design software became affordable to regular companies and small engineering firms, industrial design changed forever. Unlike a hazy human memory or hand-drawn diagram, a design file is locked down, its knowledge firmly and exactly recorded. A design file can be instantly sent halfway across the planet. Its recipient could be a stranger who without being briefed could study the captured knowledge, save it, and tinker with the original object's design or perhaps 3D print the object depicted.

Design software automates the tedious aspects of the design process. Word processing software cut out the wasted time and paper that were associated with typewriting documents. Design software eases the pain of design iterations. A designer can quickly iterate a design on screen and then go back to an earlier version. It's easy to experiment with surface texture and color by copying and pasting from one part of the design to another. Design software keeps track of every design revision and can store more data points than a human can remember.

Do you remember learning how to sketch a cube using pencil and paper? If you kept the cube's design simple, pencil and paper worked just fine. However, if you wanted to experiment with your design and add some flair, things got messy.

In primary school, I remember scrubbing at the paper with my eraser if I made a mistake or wanted to change my cube's design. If I decided too late that its corners should be drawn in fresh, bright yellow pencil instead of gray, I would have to start again with a brand new sheet of paper and a yellow pencil. Even simple design iterations were expensive in terms of paper, time, and frustration. If my teacher had asked me to add surface textures to my cube, or to sketch in a few inner tunnels, these simple additions would probably have been more than I could deliver.

My first real brush with design software was writing code for graphics design in high school. In college I found a part-time job helping a Dutch manufacturer create simple sheet metal parts on a computer. The fact that he had to ask a geeky freshmen to help him out on this shows you how few professional engineers were using CAD in those days. My classmate Guy Shaviv and I wrote a simple program that allowed an engineer to sketch out the outline of a sheet metal part, as if it were made of paper—disregarding its actual physical thickness.

The thinking was that the production floor people would enter the actual thickness of the part later, when the physical stock actually arrived on the plant. In response, the engineer's design would automatically adjust. It was a simple idea—a far cry from today's real-time, photorealistic, fly-through animations. But my faculty advisor, Moshe "Shefi" Shpitalni, had the foresight to realize that CAD wasn't just about improving the design itself.

Design software can communicate seamlessly with different divisions in a manufacturing plant. Our program allowed the manufacturer to relax its procurement criteria and purchase a much broader range of raw materials than it could with a traditional "fixed" blueprint. The software was a huge success and eventually sold for $4 million (of which my buddy and I got 2.5 dollars per hour).

Later, when I was in my early 20s and serving as a naval officer, our unit had a special blueprint division. This division was housed in a well-lit room where trained draftsmen and women toiled over slanted drafting tables drawing detailed blueprints of ship's hulls, engines, and other parts.

My commander and I lobbied to introduce 3D design software to the naval engineering unit. Eventually, we succeeded, but it wasn't an easy sell. Draftspeople had invested years in learning their trade, and it was a courageous few—typically the new recruits—who were willing to start over again learning how to use computer-based design tools. However, eventually even the top brass came to realize that design software was not only more efficient, but design iterations aside, it had built-in capacity for making calculations and predictions that no human draftsperson could touch.

Naval architects could quickly sketch out a shape for a water tank on their computers. Then in response to information from the logistics divisions, quickly digitally change the volume of the water tank. They could virtually "weigh" the tank and determine its manufacturing cost. Most importantly, the computer could calculate the effect of the water tank's shape and size choices on the ship's stability at high sea—something that would take an expert more than a week.

The great thing about design software is that it renders the physical world digital, injecting all the benefits of automation into the design process. The not-so-great thing about design software is that it still isn't, even today, capable of digitally capturing the full essence of physical objects. Computers deal well with predictable sets of a finite number of elements—for example, calculating all the different possible combinations of chess moves two players can make. Yet the physical world doesn't break down easily into a set of finite possibilities.

## Today's design software

Design software and 3D printing technologies are together leaping forward and changing the way people design and make things. However, their relationship has been largely one-sided. 3D printing grew up reliant on design software. Yet design software did not grow up reliant on 3D printing. In fact, design software is only now starting to take 3D printing seriously as a viable medium for design.

There's only so much a design file can tell a 3D printer. If we're ever going to 3D print a "real" rock, fully functioning robot, or new kidney, design software will need to raise its game. Most mass-produced objects are made in separate pieces and assembled rather than in a single complicated piece. Design software can't map what's below an object's surface. Hence, an object's innards remain beyond the reach of what a typical computer (and design software) can handle.

Image courtesy of Gonzalo Martinez, Autodesk

**3D printout of a solid model of a turbojet engine modeled using Autodesk Inventor**

Today there are two major categories of design software available for 3D printing. The first category of design software, called "solid modeling," is used by engineers and industrial designers. It offers users a ready-made library of cubes, cylinders, spheres, and other standard physical shapes that can be cut, stretched, and combined with a few clicks of the mouse. The ready-made collection of shapes is intended to jump-start the design process. From there, users tweak and adapt these basic building blocks of pre-set shapes into a unique design.

The second major type of design software is called "surface modeling." Surface modeling software got its start with cartoon animators and, more recently, video games and graphics companies. Surface modeling software works wonderfully in cases when a cartoon character or imaginary world can't be adequately captured by retrofitting an out-of-box provided library of basic shapes.

## Designing machine parts: Solid modeling CAD

Solid modeling software was born from industrial design and manufacturing. Solid modeling design packages are efficient and offer a user a built-in, ready-made library of core shapes and standard machine part geometries that can be quickly customized and combined. For example, two cylinders could be combined together to make a mallet, and a third cylinder could be subtracted to create a hole for hanging it on the wall. Because solid modeling software represents shape as a collection of volumes, most 3D printers today can "speak" to these types of design files fluently.

Solid modeling design software has years of manufacturing and design experience built into it, and it shows. The software "speaks" in an old-school machine-shop vocabulary using words such as extrude, drill, and chamfer. It shows design operations in recognizable realistic form, for example, as a drilled hole, a milled notch, or a rounded-off edge.

Solid modeling software took off in the 1990s when computing power increased to the point where the design software could "remember" iterations of previous designs and allow users to skip back and forth between them, undo changes, and retrospectively change dimensions. We take it for granted now that our computer will remember each version of a document, spreadsheet or design file, but it wasn't always so convenient. When design software finally gained this capability, it took the CAD world by storm. Parametic modeling enabled designers to keep editing and experimenting with their design—no pencil, erasers, and torn-through sheets of paper needed.

Design software has streamlined workflow, improved communications and enabled companies to store design knowledge. Before design software became a critical part of the manufacturing process, designers, engineers, and manufacturers worked in loosely connected but separate organizational compartments. These days, a solid modeling tool can tell its users whether a proposed design for a new gadget will come out right when forced into an injection molding machine using a particular type of green plastic. Design software can help designers cut down the number of separate parts to save on manufacturing and assembly costs.

Today's commercial CAD packages can be knit into a company's supply chain. If engineers want to change the size or material composition of a key engine widget, they can check the change against the company's inventory to compare the new version against existing widgets. If a new widget is added to inventory, whether it has an effect on other components will be recorded and the entire supply chain will be notified and updated.

## Drawing onscreen characters: 3D computer graphics software

If solid modeling grew up among engineers, surface modeling design software grew up in the midst of animators and illustrators. Surface modeling software originated in the land of entertainment—the cartoon, movie, and video game industries. Today scientists use 3D modeling software to create models of DNA structures or chemical compounds. Architects and landscapers create beautifully detailed models to sell their design projects to potential clients.

If a computer needs x, y, and z location coordinates to keep track of an object's shape, how does it manage to capture the details of an animated character or a complicated molecule? Surface modeling captures the world by digitally "wrapping" shapes in a virtual fishing net made of regular polygonal geometries. Sometimes called *polygonal modeling*, each polygonal shape that makes up the virtual fishing net corresponds to a data point on a theoretical grid.

Each data point is stored by the design software so the designer can work with it. Most 3D models today are built from triangular surface mesh because it's quick and a computer can render this information easily. However, no matter how finely detailed a coating of curved shapes, the triangles are ultimately planar, not curved.

The upside of surface modeling software is that it's great for depicting the variety that's present in our universe. Surface modeling software can depict beautiful detailed imaginary worlds in movies and video games. Unburdened by the limited set of block and spherical primitives, and with its emphasis on showing surface-level action rather than how machine parts fit together, surface modeling software is a graphic designer's blank slate. It's a starting point upon which an artist can fill in more details such as surface shine, textures, realistic-looking skin and hair, and wilderness.

But surface modeling software has its own burdens. Graphic animations need to move smoothly onscreen and look realistic. If light shines on a character, it must flow with the character's movements. Background must pass by at a correct speed. Keeping track of these tiny details is process called "rendering."

High-speed rendering makes for highly realistic video graphics. Rendering relies on algorithms built into the design software that are run in real time to move the action along. High-speed rendering makes surface modeling software into a veritable hog when it comes to consuming computer resources. It's possible to 3D print designs created with surface modeling software (for example, 3D printing one's video game avatar is a popular application). However, as we mentioned earlier, just printing a surface shape isn't enough. To adapt surface-modeling designs for 3D printing involves some extra steps.

In reality, most modern design tools can do both a bit of solid modeling and surface modeling. It's fairly easy to convert a solid model into a surface model by generating a virtual fishing net, or mesh, for the design object. However, going from a mesh into a solid model is more challenging. It's relatively easy to go from music notes to MP3 sound file, but the other way round is difficult. Many scientific careers have been spent trying to solve this problem, but it remains a computationally hard nut to crack.

## Design it or scan it?

Optical scans are an increasingly popular way to capture physical objects in digital form. Not too long ago, scanning something meant scanning a printed paper document or photograph to turn it into a digital file. Today, people scan building facades for an architectural project or use an MRI to scan an aching elbow to look for signs of a sprained ligament.

Scan data also captures the shape and dimensions of the physical world in a set of three-dimensional coordinates. Scan data lies somewhere between the primitive shapes of solid modeling software and the imaginary fishing net that wraps up digital objects in surface modeling software. Imagine if you doused yourself with glue and rolled around in a deep pile of multi-colored confetti. When you stood up, confetti would stick to you everywhere. Your body's surface would be snugly encased inside a dense, colorful shell of tiny paper dots.

Now imagine that someone watching this spectacle had the patience to meticulously record the precise location of every sticky, individual piece of confetti on your body's surface. Maybe on his first attempt, your scribe would patiently record each confetti's location descriptively, for example, as "one red confetti on tip of nose." After laboriously writing it out this way, your scribe would realize a more efficient way to capture the details of your confetti-covered self in all its colorful glory: to jot down the individual location of each tiny piece of paper according to its precise location in space, or by its x, y, and z coordinates.

That's essentially how scanners capture physical dimensions of things, as a surface coating of digital confetti. Each piece of digital confetti represents a data point. Each data point contains information about where each tiny dot sits on your body's surface in three-dimensional space, documented as a set of x, y, and z coordinates.

Another name for the digital confetti that coats our surfaces is a "point cloud." Most scanners capture a point cloud digitally and feed that data to a computer. After scan data is uploaded into design software, to make sense of its collection of location coordinates, the design software performs a series of rapid calculations to convert the point cloud into a surface mesh, sometimes calculating missing data points to fill in a surface gap.

3D printing and point clouds are a match made in heaven. Scan data opens up a new universe of design possibilities and unleashes the true disruptive potential of 3D printing. Scanning is useful for capturing the geometry of objects for which no design file exists, for example, natural objects such as plants, animals, and people, anatomical models, as well as of inanimate objects like rocks and even landscapes. Scanning is also useful for capturing the shape of synthetic objects when the original CAD file is unavailable or never existed, such as archeological objects and broken parts.

Scan data is, in my opinion, the bridge that's going to span the gulf between the analog physical world and the binary digital world. Scanned and reproduced physical objects are where the line begins to blur between original and replica, between copyrighted object and derivative work. Scanned data, once captured in a design file, can be edited, replicated, and copied. Someday we will edit the physical world as easily as we edit digital photographs.

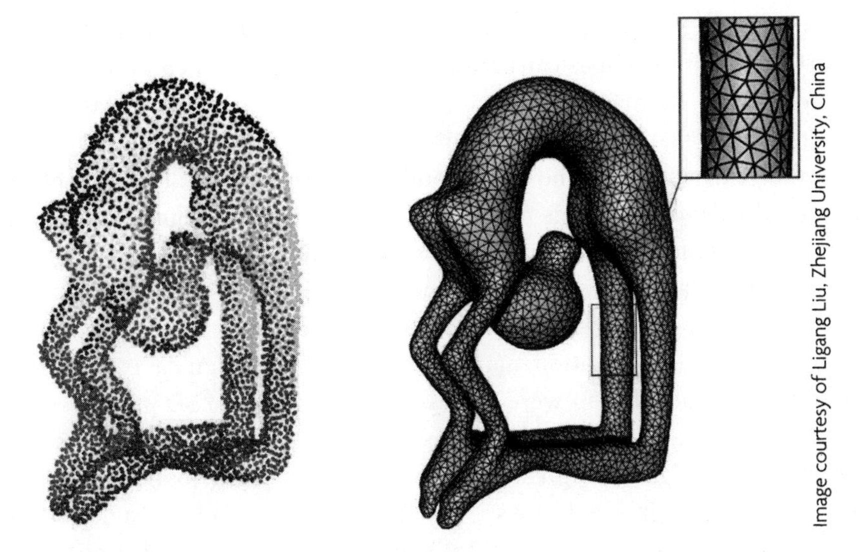


**Point cloud data, and a corresponding surface mesh**

The bottlenecks are computing power and the lack of algorithms intelligent enough to "fill in the gaps" to complete the details of a digital point cloud. The computer doesn't necessarily understand what it scanned, so you can't just scan a vase and ask the computer to make the vase wall a little thicker, because the computer wouldn't know where exactly the wall begins and ends and which direction is "thicker." Like other digital software tools, optically scanned data doesn't capture an object's insides. This is changing thanks to improving medical imaging technologies such as CT scanners, MRIs, and ultrasound.

## Why design software can't keep up

Modern design software is still dogged by its roots, by the fact that it grew out of manufacturing and animation fields that only recently started 3D printing. The same design tools that were intended to deal with limited amounts of computing power and save time, money and improve knowledge transfer, ironically, also place limitations on what can be 3D printed. As a result, a design file doesn't depict the detailed insides of physical things (at least without a lot of additional custom work). Nor can design software graphically model and predict the behavior of complicated blends of different materials.

For example, solid modeling software suffers from the fact that its standard library of primitive shapes can't always be tweaked and edited into irregular geometries. A 3D printer, however, can fabricate unusual and irregular shapes that traditional manufacturing machines could not. Therefore, a lot of design potential remains left on the table. Solid modeling software isn't capable of meeting the demands of this new and largely untapped design space. As 3D printing technologies continue to improve, traditional solid modeling will become out of date, a powerful but somewhat crude design tool.

The surface modeling software used by animators and video game designers shares a similar limitation, namely the absence of design data to describe an object's insides. If you were to design and attempt to 3D print a colorful and elaborately shaped teapot covered with amusing illustrations, the printed teapot might look great on the outside, but would not be functional. Because your design file didn't specify the shape of the inner cavity of the teapot, nor that its spout needed to be hollow and its lid to fit tightly yet come off, your 3D printed teapot would lack any sort of inner structure and would not be usable.

Even the most highly detailed, well designed 3D graphics designs can't guide a 3D printer through the process of printing anything below the surface. Given where surface modeling software got its start, this limitation makes sense. Cartoon animators never needed to 3D print their "design files." Some specialized software is available that can "guesstimate" the shape of what lies under the surface and fill in the missing details, but that process is often prone to errors and frustrations.

Printing beneath the surface is just one challenge. Another challenge lies in the fact that 3D printing technology is capable of fabricating objects so complicated their design involves more data points than even present-day computing power can handle. For example, imagine that you wanted to design and 3D print a shirt composed of millions of tiny links made of a blend of plastics—some hard and some soft. The completed printed fabric would be elastic in feel to conform precisely to your body.

With a solid modeling program, you would find defining a single ring of chain mail to be quick and easy. However, making and interlocking millions of these tiny rings would be extremely tedious and time-consuming. If you had to specify a different plastic for each specific chain mail link, the computer system would be brought to its knees. Even if a computer could keep track of all the different types of plastics to be used, 3D modeling your completed chain mail design would be far beyond the capacity of any traditional design tool today.

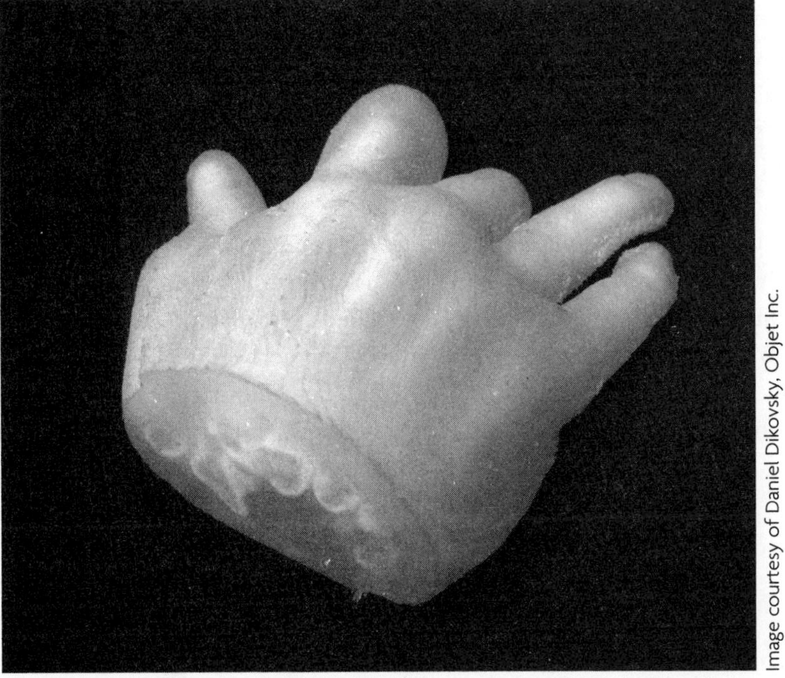

Image courtesy of Daniel Dikovsky, Objet Inc.

**A CT scan of a hand that was directly 3D printed without the scan data being converted into a surface mesh or STL file**

3D printing is moving into multiple materials, another arena where design software isn't ready yet to follow. Being able to print in several different materials at once enables designers to combine materials in entirely new ways that were not possible using traditional manufacturing. Not only does multi-material printing make it possible to produce complicated, multi-part objects in a single print job, it allows the creation of entirely new types of materials, such as a graded material that gradually transitions from one type to another.

The challenge, however, is that design software isn't yet ready to "think" about multimaterials. It's one thing to model a gear in steel or titanium. It's entirely another to model a gear that's created of a blend of light titanium on the inside that transitions into hardened steel on the outside. While some 3D printers can already fabricate such a complicated multimetal gear, there isn't yet design software that can enable this task.

Another difficulty arises when different types of materials are combined in novel ways. Sometimes material properties change dramatically as the printed shape of an object changes, adding an additional layer of complexity that today's design software isn't equipped to handle. The ability to combine several different types of materials into a single printed physical object promises vast new design possibilities, but until design software evolves to meet the challenge, these possibilities remain unexplored.

When a designed part has complicated and intricate features, is made of multiple materials, or involves millions of detailed and different design surfaces, its design file must gobble up large amounts of memory. This is why modern-day design software can't yet accurately depict a detailed model of a flesh and blood human hand, complete with cells, nerve endings, and vascular structures.

## What you design is not (necessarily) what you print

The conversion from design file to printable object is where the long, one-sided relationship between CAD and 3D printing becomes apparent. In response, people who make and work with 3D printers have devised ways to help design files print out as planned. Some software tools such as Materialise Magics and Netfabb act as "repair" tools to help users find out what's wrong with their design files.

We wanted to briefly mention another software-related aspect of 3D printing: STL files. (Remember John T. Lee from ABC Imaging?) To prepare clients' design files for printing, John converted the files into STL format. STL files play a critical role in the transition from design file to 3D printer.

### STL: The current standard

Standards and file formats are the lingua franca of technology, a critical foundation of interoperability. For example the MP3 file format enabled everyone—including makers of music players and consumers—to swap, sell, buy, and download music files, making possible today's music industry. Similar industry-standard file formats are the commercial heartbeat of other forms of media: all digital cameras save photos in JPEG format, which is compatible with laser jet printers and web browsers.

The world of 3D printers has its own industry-standard file format called Standard Tessellation Language (STL) file format. The STL file format got its start in the 1980s, an era when 3D printers were brand new prototyping tools and design software and computers were weak, tiny versions of what they've become today. Like today's design software that's now limited by its built-in inefficiencies, the STL file format was designed to simplify the transfer of design files to budding 3D printers.

To digitally "slice" a design file into a 3D printable format, early STL files had to account for the fact that a 3D printer could only handle so much physical detail. Computer memory was limited and expensive in those days. Therefore, the fact that the STL file format removed some design details was ideal since that conserved computing power. For example, a design file can contain color information and other design niceties that the STL file's job was to strip away. A typical printer needed to process only the triangles that touched the current layer and could temporarily ignore the rest until the next layer was due to be fabricated.

Fast forward three decades, and STL files remain, yet their original benefit has become a limiting factor on the design possibilities for 3D printing. If 3D printing is going to fulfill its potential, the STL format, as valuable as it has been for decades, needs to be honorably retired. Design software is improving and so are 3D printers.

Today's design software routinely handles design files that involve billions of location coordinates or intricate mesh lattices. In tandem, the best 3D printers today can keep apace and are approaching printing resolutions of 1 micron. However, the STL file, the vital bridge that spans these two technologies, can't keep up.

## AMF: The new standard

One possible way to replace and upgrade the STL file is with a new XML-based standard, the Additive Manufacturing Format (AMF). Full disclosure: I co-authored the AMF standard so, of course, I'm a fan. I worked on the AMF standard with a group of 3D printer manufacturers, CAD software vendors, and expert users. We teamed up under an international organization that manages the development and implementation of technology standards, the ASTM.

AMF maintains the surface mesh structure of the STL format but has added capabilities to reflect advances in design software and 3D printers. For example, the AMF file format can handle different colors, different types of materials, the creation of lattices, and other detailed internal structures that are one of the huge benefits of additive manufacturing. Curved triangles can be used to describe curved surfaces more accurately and more compactly than the planar triangles used by STL.

The AMF standard was officially approved by the standards body in May 2010, but the ultimate test of any standard is its adoption. At the time of writing of this book, it has yet to be adopted by 3D printing vendors. It may take years. We're stuck in a chicken-and-egg paradox: CAD vendors and 3D printing companies are waiting to see whether anyone will gamble on the new format and abandon the old but tired warhorse STL.

## The next generation of design software: digital capture

If the first mainstream generation of design software rode in on the tidal wave of desktop computing and traditional manufacturing, the new generation of design software will ride in on the tidal wave of digital capture, or "reality capture." The marketable future of design software is to make reality capture something everyone can work with. Amongst software companies, the race is on to capture the consumer markets. Autodesk, the same company that in the 1980s launched MicroCAD, the word processor for drawings, hopes to again be at the crest of the next big wave of change. Autodesk's answer has been to create 123D, a suite of no-cost design tools aimed at children and consumers.

Other CAD companies (SolidWorks, PTC, Rhino, and SketchUp are but a few) are building modules for consumer designers. Yet like any business that has captured its place in a mature field, software design companies are faced with a choice that is difficult to bridge. On one hand, the company must adapt its organizational resources to capture this new market opportunity. On the other hand, the company must continue to maintain its core productivity engine that still serves 99 percent of traditional manufacturing (and brings in the company's revenue).

"The future of design software is to make it easier to move back and forth between reality and software, software and reality," said Gonzalo Martinez, a vice president of Autodesk's consumer products division. "Our goal is to speed the process. Now we have tools that can capture an object's physical details in as many as a billion data points. Our goal is to make it easier for people to manipulate that data to understand from a human point of view, to answer the question, 'What is that object?'"

"We're exploring new ways to design and fabricate things" said Gonzalo. "Last year we built a state of the art fabrication lab. Now we have the biggest collection of 3D printers on the west coast," he joked.

If capturing the design details of physical things becomes a quick and painless process, then everybody can become a designer. Once 3D printing becomes as ubiquitous as 3D printing, everyone can become a manufacturer. "My 11-year-old son will design a complex object that once took me 3 years to learn how to design," laughed Gonzalo.

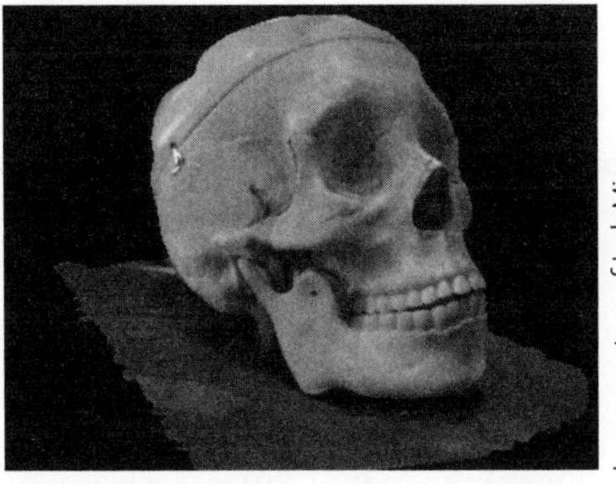

Image courtesy of Josh Mings

**This 3D printed skull was captured in digital form with 123D Catch.**

Gonzalo continued, "The long-term goal is to generate parametrics [shape libraries] from point cloud data. This is good for reverse engineering and to speed up the design process if we want to duplicate the object exactly in digital form or change it. In a nutshell, the next generation of design software will bring fast reality into the computer."

# 7 Biopringing in "living ink"

Anthony Atala, a researcher at Wake Forest University, caused a sensation when he appeared in a TED talk in 2011 and gave what many people mistook for a demonstration of how to print a living human kidney. Naturally, since 90 percent of the patients on the organ donation list are waiting for replacement kidneys, people got very excited. After the ensuing confusion was sorted out, it turned out that 3D printing live kidneys was still in the early research phase. The "kidney printing" was actually a lab experiment involving the 3D printing of kidney-like tissue that was capable of filtering blood and diluting urine.

Atala's TED demonstration had the effect of raising peoples' awareness of the possibilities inherent in 3D printing body parts. Atala, long an evangelist of regenerative medicine and one of the pioneering researchers of bioprinting, remains optimistic. In an interview with a newsletter from a major financial firm, Atala said that "there is no question that someday, perhaps in the span of a generation, you can have a heart made out of your own cell tissue. Isn't that amazing?"[1]

## The printer of youth

William Shakespeare called old age a "hideous winter." Many cultures tell stories of a mythical Fountain of Youth which gives people the gift of eternal youth. A medieval novel described the Fountain's therapeutic power to turn old warriors into young ones.

> *The old warriors; more than forty-six bathed in it and when they came out they were age thirty and like the best knights. Then the other old men . . . said see how old and bent we are? We have lived more than a hundred years and now you will see us in another guise.*

> *They entered the fountain and bathed four times as prescribed. They left the fountain rejoicing, and when they returned to Alexander he could hardly recognize them, so young they were.*[2]

Despite centuries of searching, no one has yet found the Fountain of Youth. In modern times, people continue the quest with plastic surgery, replacement joints, new heart valves, miraculous vitamin supplements and skin creams. Maybe 3D printing technologies will finally end the search. Printed on-demand body parts will help people who need an organ transplant, or have failing joints. People with disposable income will order custom printed body parts optimized for a beloved recreational activity. The Olympic Committee in the year 2072 will struggle to to decide whether athletes with bioprinted organs should be banned from the Games.

3D printed body parts are still the stuff of fiction. Today the science of printing living tissue is just beginning its long ascent up a hypothetical "3D printed Ladder of Life."

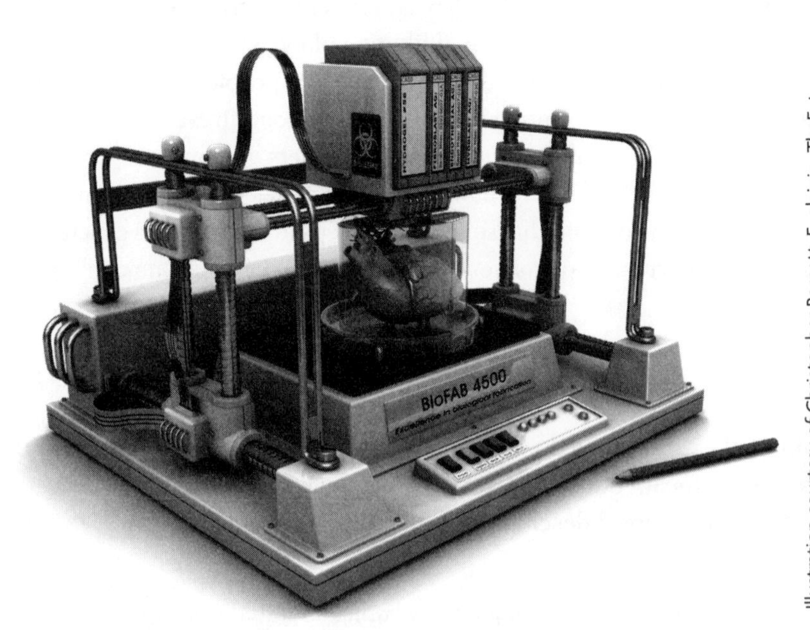


Illustration courtesy of Christopher Barnatt, ExplainingTheFuture.com-Bioprinter concept created by and copyright © Christopher Barnatt, ExplainingTheFuture.com

**The holy grail of bioprinting is to 3D-print a functional organ from cell mixture and biomaterials.**

Imagine a tall ladder of body parts arranged in order according to their complexity. Inanimate prosthetic parts would sit on the lower rungs. On the middle rungs would be simple living tissue such as cartilage and bone. Above simple tissue would sit veins and skin. Just below the top of the ladder would reside complex, critical organs such as the heart and liver, and the brain. Finally, perched atop the Ladder of Life would be complete living creatures, or perhaps someday, fully functioning synthetic life forms. Today, 3D printing technology has already placed its feet on the lower rungs of this imaginary ladder, aspires to the middle rungs and dreams of someday attaining the top rung.

## 3D printing the ladder of life

In a commencement speech at a Delaware High School, Joe Biden described a glowing future, when "using 3D printers, we're going to be able to restore tissue after traumatic injury or a burn, restore it back to its original state. It's literally around the corner."[3] Biden's bold claims are just that: bold claims. However, he's not entirely off the mark.

Let's start with the lowest rungs on the ladder—inanimate replacement "body parts" such as dental crowns or artificial limbs. The first wave of commercially available 3D printed body parts are already out there walking around inside the bodies or regular people, perhaps even in yours. Non-living prosthetics such as 3D printed bone implants, dental crowns, contact lenses, and hearing aids reside in thousands of humans worldwide.

Phil Reeves, the managing director of Econolyst, a consulting company dedicated to the 3D printing industry, estimates that today there are "ten million 3D printed hearing aids in circulation worldwide."[4] Invisalign braces—3D printed, custom-made, clear disposable plastic braces that hide over a patient's teeth to pull them into alignment—have been a tremendous commercial success. There are an estimated half to three-quarters of a million 3D printed dental implants travelling around in people's mouths right now.[5]

Like high-end 3D printed titanium airplane parts, 3D printed human body parts represent the ultimate in small batch, direct to digital custom manufacturing. The process for teeth, hearing aids, and braces is similar: the body part in question is scanned. The scan data is sent to a special lab where it's adjusted into a viable design file. The design file is 3D printed into soft rubber, hard and shiny ceramic, or soft flexible transparent plastic.

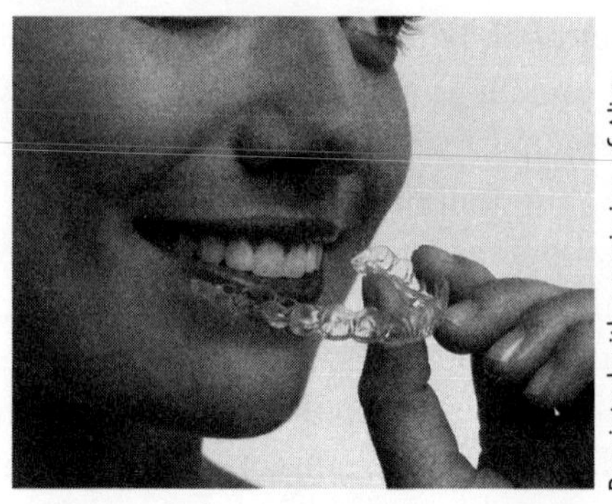

Reprinted with permission of Align Technology, Inc.

**Approximately 50,000 custom Invisalign braces are 3D printed every day.**

Today's first wave of 3D printed body parts are made of a single material such as metals, ceramic or plastic. They make sense commercially since their market value rests on the fact they fit closely into a uniquely shaped body. Their custom shape, small batch production, and the fact they offer their manufacturer no economy of scale makes them perfect (and profitable) candidates for 3D printing. Getting past regulatory barriers is relatively simple. Unlike living tissue or medications, inert bodily "bolt-ons" involve fewer urgent health risks and their side effects are more predictable.

How about an artificial limb that has sex appeal? Today, about two million people in the United States have artificial limbs according to a national advocacy group, the Amputee Coalition. Most modern artificial limbs are still essentially unchanged from those fitted onto soldiers returning from the devastating battlefields of World War II. An artificial hand is made of metal, and its ability to grip small things is provided by a pincer-like set of hooks.

Bespoke Innovations, a small San Francisco-based company recently acquired by 3D Systems, designs and 3D prints custom prosthetic limbs. "The way most artificial limbs are made hasn't changed much over the years—you take a piece of foam, shave it into a rough approximate of a person's leg, then make a mold and stamp it out," said Scott Summit, a co-founder of the company.[6]

The science of artificial limbs is making a big leap forward thanks to improved medical imaging technology, better design software, and improved materials. 3D printing offers amputees and doctors a previously unheard-of level of customization. Bespoke's 3D printed limbs are designed to fit exactly the shape of the wearers' bodies and lifestyles and to appeal to their sense of style.

Bespoke's process begins with scanning both a patient's "sound side" leg and their current prosthetic leg. The data is modeled into a computer design file and their "sound" leg is superimposed onto the digital image of the artificial leg to make sure their new custom limb gives them their body symmetry back. After customers have selected their own unique design, or fairing, Bespoke 3D prints it out.

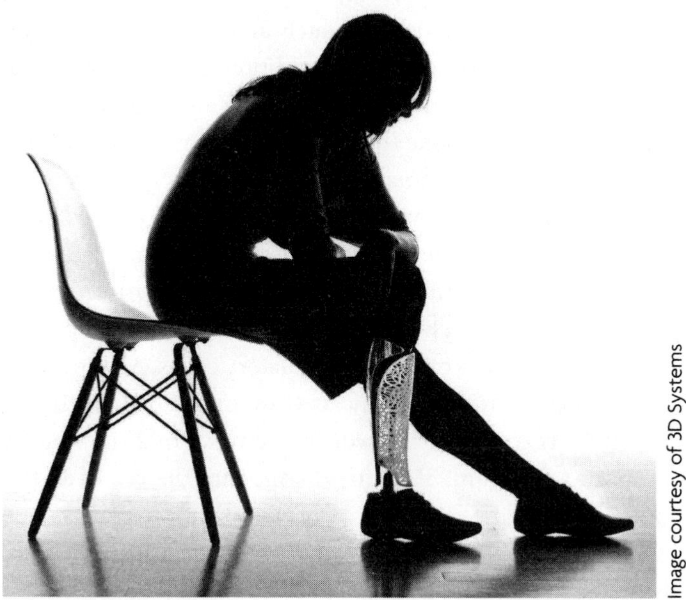

Image courtesy of 3D Systems

**A stylish artificial limb**

Bespoke's online tool called the "Configurator" allows customers to explore a range of design styles, including leg patterns, materials plating, and finishes that give its wearer the same social cachet offered by a racy designer motorcycle or unique tattoo. The company named its artificial limbs "fairings" after the specialized coverings used in sleek, high-end motorcycles.

Bespoke's website describes its offerings as follows. "Bespoke Fairings . . . not only return the lost contour, they invite an expression of personality and individuality that has never before been possible." Pictures of people wearing Bespoke's custom-made artificial limbs demonstrate the shift in mentality from the days that people used to hide their artificial limbs. Bespoke's customers wear their artificial limbs proudly.

If we return to the 3D printed Ladder of Life and ascend another rung, we leave behind prosthetics and reach the next rung, 3D printed bone implants. 3D printed bone implants and artificial joints are still considered exploratory medicine. At the time of this writing, 3D printed, custom-made artificial joints made of titanium are available to patients fortunate (or courageous enough) to receive advanced and/or experimental medical care.

Most standard bone replacements are injection molded if they're polymer or cast in metal if they're made of titanium. The same limitations that apply to making any plastic or metal machine part also apply to making bones. For example, separate bone parts must be molded separately and then assembled later. Freshly molded bones demand precision cooling conditions so they won't shrink or distort and will remain clean. Overcooling can make a new polymer bone brittle; undercooling will result in a bone that's too soft and will smear while handled.

3D printed titanium bone implants have received regulatory approval, but printed polymer bone implants have not. When polymer printed bones receive regulatory approval, they offer new possibilities since polymer has special properties that titanium and ceramic lack. For example, a 3D printed polymer bone could be infused with bioactive bone growth additives and active pharmaceutical ingredients such as antibiotics or anti-inflammatory drugs. A 3D print head could spray droplets of these bioactive chemicals with unmatched precision.

One dramatic surgery made the worldwide news in 2012 when a team of surgeons inserted a titanium, 3D printed bone into the jaw of an 83-year old Belgian woman with oral cancer. The process began when the medical team took a CT scan of the woman's jaw. A medical design company, Xilloc Medical, adjusted the CT scan data into a printable design file and used computer algorithms to add thousands of irregular grooves and hollows into the jawbone. This way, the woman's veins, muscles and nerves could knit themselves more quickly into the new jawbone to fully integrate it into her body.

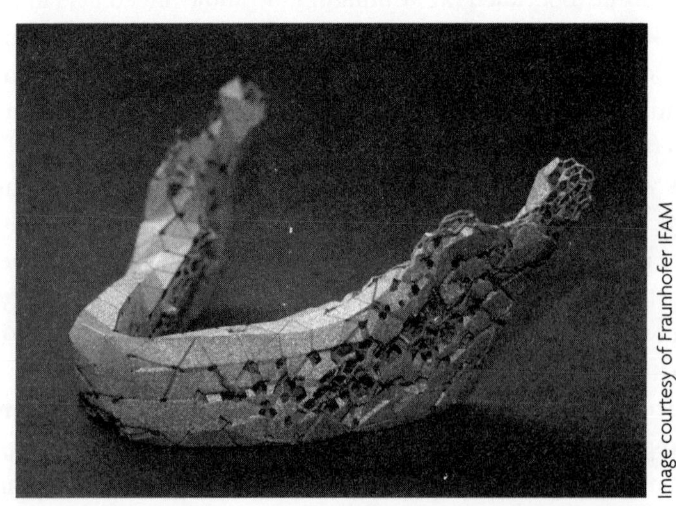

Image courtesy of Fraunhofer IFAM
Human lower jaw with biomimetically graduated cellular structure, made by SLM (metallic bone foam) (c) Fraunhofer IFAM.

**An individual jawbone implant "baked" from powdered metal**

In the final step in the process, the titanium jawbone was printed by a Belgian company called LayerWise that specializes in high-end, medical-grade 3D printing technology. The smooth, gleaming new titanium jaw was printed by shining a laser into titanium powder, fusing 3,000 meticulously laid layers. Finally, the printed bone was coated with ceramic. Hours after the surgery, the woman spoke and sipped soup.

## Tissue engineering

The term *bionic* apparently was derived by combing the Greek word *bios*, or living, with the word fragment *onic*, from electronic. The word *bionic* was made famous by a popular 1970s TV show that featured Steve Austin, the Bionic Man. Steve's significant other, the Bionic Woman, had her own successful spinoff and periodically made guest appearances on the Bionic Man.

The Bionic Man and Woman had superhuman strength and laser-keen senses thanks to the insertion of several very expensive artificial body parts. In nations that broadcast these shows, backyards and playgrounds would empty out at regular intervals as kids disappeared to watch TV. I remember packs of my friends and I running in extreme slow motion, making electronic clicking sounds while we performed bold physical stunts or made daring escapes from behind one another's enemy lines.

Bionic parts sounded promising several decades ago. In the future, the notion of bionic body parts will seem crude—prone to malfunction, poor fit, and unable to evolve—similar to the limitations of today's artificial replacement hips and knees. In the future, bionic parts will be replaced by custom-designed 3D printed tissue.

If plastic is the favored raw printing material in industry, stem cells will be the favored raw material of bioprinting. The more regenerative medicine advances, the more it comes back to nature. The future of tissue engineering lies in 3D printing stem cells into precise configurations and letting them do the work growing the living tissue.

Stem cells are the raw clay of the human body and are much more skillful at making body parts than we are. Stems cells are unspecialized, meaning they have not yet committed to a particular career path in terms of which sort of bodily cell they will grow into. Since stem cells can be pushed to differentiate into any one of the approximately 210 cell types found in the human body, stem cells are pure gold, medically speaking.

The first stem cells identified in the 1980s were extracted from the tissue of unborn human fetuses, triggering emotional debates about medical ethics. Since then, researchers continue to uncover more stem cells, including some scattered around different parts of the adult human body called "somatic" or "adult" stem cells. More recently, some differentiated cells have been shown even to be able to revert back to their pre-differentiated state.

Columbia professor Jeremy Mao 3D printed new hip bones in lab rabbits and seeded them with stem cells. First, Mao and his team removed and imaged the hip bones of lab rabbits. They converted the images to a working design file, 3D printed the hip replacements out of artificial bone, then sprinkled the artificial bone with the rabbit's stem cells and surgically inserted the bone back inside the rabbit. By the end of 4 months, all the rabbits were walking freely, some even placing weight on their new hips a few weeks after surgery.

Mao and his research team designed the replacement bone with tiny curving inner micro channels that encouraged the stem cells to creep along the implant's surface and help the rabbit heal more quickly. In other experiments, Mao's team printed out a replacement incisor for a lab rat, sprinkled it with stem cells and implanted the replacement back into the jaw. Nine weeks later, thanks to the incisor's ideal shape and stem cell infusion, new ligaments and bone embraced the artificial tooth.

Similar to the approach used by Dr. Mao's team at Columbia University, researchers at Washington State University 3D printed bone using a fine misty spray made of calcium phosphate, silicon, and zinc powder. The fine spray droplets created layers 20 microns thick (half the width of a human hair). The printed bone was sprinkled with immature human bone cells. The method worked, as the immature bone cells thrived in their new environment and eventually grew into mature, living bone tissue.

## Stem cells, bio-ink, and bio-paper

A medical dictionary defines tissue as "an aggregation of similarly specialized cells which together perform certain special functions." Our bodies are made up of different types of tissue, from adipose tissues that form our fat cells, to cartilaginous tissue that form cartilage to cushion our joints, to nervous tissue made up of complex networks of connected neurons. Soft tissues hold their shape thanks to an internal supportive infrastructure.

When we can print living cells and make them grow into living tissue, we will have successfully ascended to the top rungs of the hypothetical 3D printed Ladder of Life. True bioprinting, as we define it, creates living tissue, not inanimate replacement parts. Bioprinting involves placing living cells into just the right location with a 3D printer, to fabricate functional, heterogeneous living tissue.

Researchers define "bioprinting" in different ways, and as the field advances, the term will likely take on even more meanings. One approach to bioprinting involves the use of a "living ink" that's a printable gel with living cells suspended inside. The special gel—called hydrogel—cushions and protects the living cells as they are pushed through the printing nozzle. Once the living ink has been printed out and laid down into the right place, the hydrogel maintains the tissue's desired structure. The living cells will secrete a substance into the hydrogel that eventually forms a supportive matrix. As the living cells continue to develop, the matrix develops into cartilage or some other type of living tissue.

Depositing the right cell type into the right location is somewhat like the process of planting the perfect vegetable garden, with each vegetable planted precisely in the right location to receive the optimal amount of light. Not all stem cells are the same. So far, nature is still much better than humans or computers at creating the perfect garden of stem cells placed into precise position.

One of the major advantages of 3D printing with living ink to make soft tissue is a printer's ability to carefully squirt cells into precise patterns and shapes. As printers get better at printing using multiple print heads, each print head can be filled with a different cell type. The result will be that one nozzle would print a different type of cell, and another nozzle would print hydrogel with different material qualities. By borrowing the concept of multi-material 3D printing and applying it to the biological world, researchers are steps closer to creating artificial tissue that mimics nature's complicated shapes, internal structures, and cellular diversity.

Cell placement is one challenge. Another challenge is making sure that cells are placed in such a way that they will eventually form the right shape. Cell location and the shape of the resulting tissue is critical for proper organ functioning. Growth factors could also be printed and added into the mix.

For example, heart tissue requires high cell density to make sure that the heart beats in a regular rhythm. If the cells seeded onto the scaffolding in artificial cardiac tissue aren't tightly interconnected, the result is an irregular heartbeat. Since a tissue design is based on a computer-guided design, bioprinted tissue made using living ink will be precisely made and its design repeatable.

Finally, another mystery humans have yet to solve is the fact that living cells need a "start" button. Today, no one knows exactly how to jump start seeded cells, even if the cells were placed into the right spot in a perfectly shaped scaffolding. Nature knows how to make an organ start working. We don't.

A team of researchers at the University of Missouri and Yale coined the terms *bioink particles* (multicellular spheroids) and *biopaper* (biocompatible gel).[7] In an article in *Nature* magazine, Dr. Gabor Forgacs, a researcher at the University of Missouri, describes his approach, "You give us your cells: we grow them, we print them, the structure forms and we are ready to go."

His team used a custom-made 3D printer originally intended to make microelectronics:

> *The printer has three heads, each of which is controlled by an attached computer, that can lay down spheroids of cells much as a desk printer would lay down ink.*
>
> *Two of the heads print out tissue cells (mixtures including, for example, cardiac and endothelial cells), while the third prints a "gap-filler" (such as collagen) that fills a space temporarily until the other cells have fused. So to make a blood vessel, for example, lines of cells are laid down with lines of collagen in the middle, which will later be extracted to make way for blood.* [8]

To understand why bioprinting holds such potential, it helps to briefly compare it to established methods of tissue engineering.

For several decades, living tissue has been made using a two-phase approach. The first step is to engineer a tissue's scaffolding from some kind of biodegradable material. To make engineered scaffolding using the traditional approach, researchers would stamp its shape from a mold, carve it, or use chemicals to etch out a porous shape. The second phase is to seed the scaffolding with living cells.

Compare this approach to just 3D printing living cells into a precise shape and letting them form their own matrix and eventually supportive scaffolding. Traditional tissue engineering techniques have helped thousands of patients gain back lost soft tissues. Yet, limitations are aplenty.

As described by researchers Miguel Castilho, Ines Pires, Barbara Gouveia and Jorge Rodrigues,

> *the drawbacks of these techniques include the extensive use of highly toxic organic solvents, long fabrication periods, labour-intensive processes, incomplete removal of residual particulates in the polymer matrix, poor repeatability, irregularly shaped pores, insufficient interconnectivity of pores and thin structures. In addition, most of these methods bear restrictions on shape control.*[9]

In other words, manually created artificial scaffolding can biodegrade and scatter cells and particles into disarray. In addition, engineered tissue built on pre-formed scaffolding seeded with living cells may not fit neatly into a patient's existing living tissue. Since human tissue comes in so many odd and elaborate shapes, it can be difficult to make a tissue scaffold that's precisely the right contours. And finally, to top it all off, it's very difficult to seed multiple cell types into the inside of an existing scaffold.

## Printing living cartilage

Cartilage is a shining example of nature's tissue engineering abilities. Inside the body, cartilage has an amazing ability to hold its shape for years, or when it's cushioning our joints, to endure years of pounding. Knee joints are cushioned and smoothed out by a protective layer of articular cartilage that prevents the bones from grinding against one another. Cartilage is what makes our ears and noses bendable, yet resilient when tweaked.

Like bone, cartilage tissue is simple tissue that is made up of just a few cell types, contains no veins, and has a relatively simple purpose. Cartilage doesn't have to digest food, obey instructions from nerve cells or respond quickly to environmental cues. Yet, fabricating even a relatively simple tissue such as cartilage is still not something the medical professional has figured out how to do well.

Cartilage is an essential tissue but unfortunately, today we have no viable way to make artificial replacement cartilage. If you've played years of squash or are a dedicated long-distance runner, you already know that once your articular cartilage is worn thin, it's just gone. Its absence in between the joints can be devastating, causing pain and osteoarthritis in millions of people who suffer from bad knees, elbows, stiff hips, and fingers.

3D printing holds promise as a method to create artificial cartilage. At Cornell University, Danny Cohen, Larry Bonassar, and I 3D printed a sheep's meniscus. First, we took an MRI scan of the sheep's knee and adapted the image data into a design file. Next, we extracted living cells from the sheep and stirred them into a medical hydrogel. Finally, we squeezed the gel mixture through a 3D printed head, in this case, a syringe. In a later research project, Larry Bonassar 3D printed real human ear cartilage whose "design file" originated in data from an optical scan of an outer ear.

Image courtesy of Daniel Cohen and Larry Bonassar

**We 3D printed an artificial meniscus of living cells from a sheep. The "design file" was a CT scan. The printed cells thrived in their new environment but the artificial meniscus was never implanted.**

There's nothing simple about the human body and creating even a simple tissue such as cartilage remains a complex process. Although it's possible to 3D print living cartilage, just successfully printing living cartilage is just half the battle. We haven't solved a critical second challenge. Our joints were designed to take punishment. Artificially made cartilage needs to be toughened up and conditioned before it's ready to be transplanted into someone's body. Therefore, cartilage made in a research lab must be subject to mechanical stress before it can be implanted.

Like an indulged child protected from his environment, artificially made cartilage that has enjoyed a privilege life inside a sheltered petri dish hasn't faced the reality of life's relentless pounding. In the absence of external stressors such as miles of swimming or pounding games of tennis, artificial cartilage remains flaccid and weak. If artifical cartilage were to be inserted into a knee, uncured, it would squish away into nothing. When Danny, Larry, and I showed our first batch of printed sheep cartilage to a bunch of practicing surgeons, they quickly ushered us out of the room once they figured out that the printed cartilage was too weak to maintain even simple sutures.

Researchers are looking for solutions to this problem. One promising method is to place artificial tissue into a bioreactor to mimic the way real tissue is pounded into maturity. To prepare bioprinted cartilage for real use, maybe the solution will be provided by advances in hydrogel materials. Another possible solution may be the creative application of other sources of stress (such as light or heat or sound vibrations) to prepare embryonic bioprinted cartilage for what lies ahead. Yes, even cartilage needs tough love.

## Printing heart valves

Cartilage may be a relatively simple type of living tissue but even some kinds of cartilage are more complex and critical than others. For example, if the cartilage in your knee or elbow is destroyed, you continue to live on (albeit immobilized and in pain). However, if the cartilage in your heart valves isn't doing its job, your risk of dying from some sort of cardiovascular disease increases by 50 percent.

There's no organ more mission critical than the heart. The heart is made up of muscle, blood vessels, and cartilage that dance together in a complicated routine that's choreographed by electrical impulses that shoot through the body. The average human heart beats nearly 100,000 times per day. In fact, heart tissue withstands its own form of pounding, at an average rate of about 80 million beats a year, about 5 to 6 billion beats in an average lifetime.

One of the most medically problematic parts of the heart are its thin fibrous valves. The human heart has four chambers that are separated by valves. Heart valves are like one-way gates that open and shut in precise time to control the direction of blood flow as it pumps from chamber to chamber. If these valves don't work properly, a patient's heart will eventually fail. The American Heart Association reports that five million Americans each year learn they have heart valve disease; defective heart valves are a common birth defect.

Heart valves are quite small, anywhere from between the size of a dime in a newborn to the size of a quarter in an adult. Blood flow must pulse hard in only one direction. If a heart valve is mechanically inadequate, it will start to slowly leak, sort of like an incompetent employee in an organization whose shoddy work eventually causes highly functioning units to slowly deteriorate as well. Valves can also cause the heart to fail if they become thickened or stiffened.

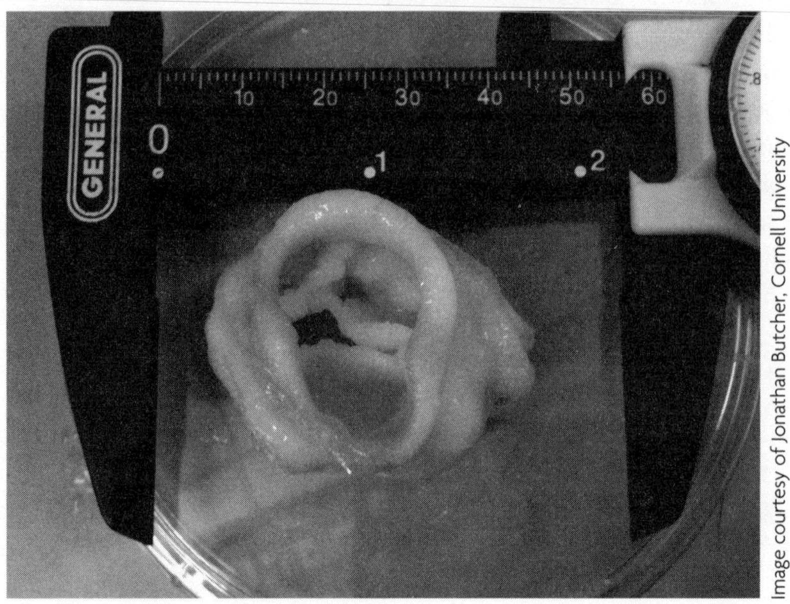


Image courtesy of Jonathan Butcher, Cornell University

**3D printed artificial heart valve**

Someday 3D printed heart valves may provide a solution. Jonathan Butcher is a professor at Cornell University and one of the leading researchers in the field of bioengineering artificial heart valves. I visited Jonathan in Cornell's bio-engineering department which is housed in a brand new white marble building. Inside, icy stone floors amplify the echo of footsteps. Otherwise the atmosphere is hushed. The reception area's oversized boxy atrium that greets visitors feels too large, dwarfing the puny humans inside.

Jonathan's office was a respite from its stark, echoing environment. Two colorful oil paintings warmed the office, a gift from an undergraduate in one of Jonathan's classes. "She oil-painted some of the heart organs she isolated from chick embryos at different stages of development," Jonathan told me. I asked him to explain the challenge of today's artificial heart valves. He said, "Today surgeons replace heart valves in one of two ways: mechanical heart valves or valves harvested from a cow or pig that are cleaned and cured like a soft piece of leather.".

Currently available valves—both mechanical ones or valves taken from an animal—suffer from serious drawbacks. The upside of a mechanical heart valve is that it lasts a long time after it's implanted. However, a mechanical valve can trigger blood clots that break off and migrate into the brain. That's why people with a mechanical heart valve must take blood thinning medicine, causing a cascade of yet another set of medical challenges, as well as occupational and lifestyle limitations.

Heart values transplanted from an animal, typically a pig, don't require blood thinners but they don't last as long. Transplanted animal valves aren't durable enough to survive for long in the active bodies of younger patients. Finally, neither mechanical nor biologically derived prosthetic valves can grow in tandem with their living host, requiring that recipients receive repeat open heart surgeries to implant increasingly larger valves.

Jonathan explained that someday surgeons will save human lives by taking an entirely new approach: 3D printing a new heart valve that's implanted directly inside a young child suffering from a congenital defect. Jonathan believes that a critical part of such a solution lies in unlocking the mystery of how the stem cells in an embryo develop into mature heart valve cells. If he can gain insight into this maturation process, Jonathan believes he will be one step closer to someday bioprinting a functioning artificial heart valve.

Jonathan's research aims to crack three parts of the bioprinting puzzle. First, he's addressing the old tough love problem. To function property, printed heart valves, like joint cartilage, need to be beaten up a bit in an incubator called a bioreactor.

Jonathan is working on methods to perfect the bioprinting of many different types of stem cells in a single "print job." To mimic nature's ability to swirl together lots of different types of cells into precise, mission-critical patterns, Jonathan 3D prints in living ink using several printing nozzles at one time. To bioprint different types of stem cells at once, Jonathan modified a Fab@Home bioprinter with multiple syringes.

Finally, since bioprinting is by definition multi-material printing, Jonathan is developing software that can choreograph the movements of several print heads that each contains a different cell type. "A 3D printer can follow instructions from a design file to print just one type of material. So we had to invent a software algorithm that enables a single design file to manage a multi-nozzle 3D printer to enable us to print several different types of cells at once," he said.

Much of Jonathan's focus is on defining the optimal shape for the deposit of bioprinted stem cells. Since cells on a heart valve must be densely packed in a particular placement to function, cell placement is critical. Stem cells are like hard-working, self-directed employees that just need the right kind of work environment. "Think of a stem cell as a worker bee," explained Jonathan. "If you can find a way to print certain types of stem cells into exactly the right location on the engineered tissue, it's like having a stem cell walk into an empty office and start to look for work to do."

As researchers like Jonathan continue to unwrap the mysteries of 3D printed living tissue, hopefully the risks of organ transplants will someday diminish. The beauty of 3D printing stem cells harvested from a patient's own body is that the patient's immune system is more likely to accept the printed replacement organ. Debilitating immunosuppressive drugs, similar to the blood thinners used by heart valve replacement patients, introduce a downstream cascade of negative side effects. Printed heart valve implants made from a child's own stem cells would be able to grow with the body and repair themselves.

## CAD for the body

No matter how skilled the designer, it's impossible to boot up your computer and pull up a "Body CAD" software program. Though some emerging types of software are developing in that direction, we're light years away from being able to design a new knee that's optimized for a female Tae Kwon Do black belt who's 5 feet, 10 inches tall, weighs 145 pounds, and is prone to repeated knee injuries of the right medial ligament.

Commercial design software has been shaped by its origins: engineering (product design) or computer graphics (animation or video games). One reason for current limitations inherent in design software tools is the fact that until recently, few people imagined that we would ever really need CAD for the body. If you think about it, fabricating new body parts will likely follow a similar design process to that of fabricating a new machine part or creating a new animated movie.

In search of more insight into the notion of CAD for the body, I journeyed west to the University of Utah, a world-class research university located in a remote, but geographically stunning corner of the world. For decades, the University of Utah has been a hotbed of innovation in digital imaging. The list of Utah alumni reads like a *Who's Who* of computer graphics: John Warnock, one of the founders of Adobe; Ed Catmull, founder of Pixar; and Jim Clark, founder of Silicon Graphics and later Netscape. Nolan Bushnell, another alum, created Pong, one of the first commercially successful video games while working at Atari.

Utah's campus in Salt Lake City nestles in the bosom of a gigantic crater-shaped hollow surrounded by jagged, mountain peaks that glow white in the winter and shimmer with lush green foliage in the summer. To someone used to the less dramatic landscapes of the east coast of the United States, Salt Lake City resembles a moonscape. Like a multi-million dollar lunar research station, Utah's world-renowned Scientific Computing and Imaging Institute (called the SCI Institute), founded in 1994, sits proudly on the edge of campus, its banks of windows looking over the surrounding jagged rim of mountain peaks.

The research at Utah's SCI institute aims to converge the fields of medical imaging, visualization, scientific computing, and Big Data research. There's something in the air in Utah that gives one a feeling that anything is possible. The state remains sparsely populated. It's a skier's and hiker's paradise, home to several national state parks and a series of scenic highways that thread through pristine stretches of wilderness.

In a peaceful office whose windows were lined with orchids, I sat down with Chris Johnson, founding Director of the SCI Institute. Chris was calm and unhurried. Chris's polite, laid back demeanor belied the fact that over the years, he has won a string of awards for his work in biomedical computing and imaging, most recently, the IEEE Charles Babbage Award, computing's version of an Oscar for lifetime achievement. Despite my sudden appearance at his Institute's front desk, he was courteous and took the time to give me a tour of the Institute's sparkling new four-story building.

I asked Chris the key question: Will there ever be commercial design software to design and improve human body parts? "Maybe," he said.

Chris elaborated, "Right now, the body is too complicated from a geometric point of view, and CAD models are based on regular geometries." Translation: Our bodies are so complex, made up of a broad range of materials and intricately shaped tissues, bones and blood vessels, that it's currently impossible to digitally capture enough detail with existing software and hardware to 3D print real, living organs.

Chris sees what he calls a "great merging" between medical imaging, big data, video game animation, and conventional computer-aided design software. Research scientists at SCI explore (amongst other things) how to capture and simulate the human body in digital form, software development that will someday play a critical role in the development of 3D printing living tissue.

Other researchers are developing computer algorithms that can skillfully stitch together scanned cross sections of a faulty organ into a single 3D computer model. A major challenge continues to be how to best manage the enormous reams of data generated by the medical imaging process.

In the larger computing industry, commercial video game designers are making strides capturing surface details and better understanding how to graphically depict the way our bodies move. The medical establishment is making progress in capturing more precise digital details about the insides of our bodies. Academic scientists are building ever-more powerful algorithms to model, predict, and analyze data collected from biological systems.

Real bioprinting—designing and editing living tissue and body parts—won't become a reality until there's truly usable CAD for the body. Medical imaging, as powerful as it can be, can only generate a design file from a body part that already exists. There's no software that can enable even a skilled medical professional to design an entirely new heart from scratch or even edit an existing one.

The problem with the human body is that it's irregularly shaped. Our bodies are miracles of geometric complexity. Each of us has our own unique body shape with lots of surface curves, different cell types and millions of minute details. Delving deeper, under the body's skin lies a biological wonderland whose complexities rival those of the galaxy, rife with mysteries.

Bodies change constantly. They're moody, prone to ever-changing states of constant flux in response to changes in our environment, our emotions and what we eat. Millions of cells grow, heal, and change on a daily basis in mysterious ways we don't yet fully understand. Cells signal to one another in ways that we have yet to decode.

Advances in traditional design software, medical imaging and data analysis will pave the way for 3D printing replacement parts for living creatures. But will we soon see commercial design software for body parts? Not yet. But we're inching closer every year. "I'm seeing a convergence between the world of simulation software, medical imaging, and CAD systems." Chris joked, "It will personalize medicine for us, which is good if you plan to need extra body parts in the future."

In industrial product design, designers are learning that as their design tools improve, nature becomes an increasingly useful source of inspiration. In body design, it will be the same. Living creatures are the product of millions of years of ruthless, iterative design cycles. Janine Benyus, an author and design visionary, said, "We've discovered again and again that biomimicry works because it offers a turnaround strategy for our species, a practical way for us to fit in and flourish on this planet by emulating 3.8 billion years of brilliant designs and strategies."[10]

## Shades of gray: Taking pictures inside the body

If there's no design software for body parts, how is it possible that people today are able to print teeth and bone replacements? How did the surgeons who performed the titanium jaw replacement manage to 3D print the exact shape of the replacement jaw? Simple. They CT scanned the patient's body, captured the image data in a file, and then massaged the image file into a file format that could guide a 3D printer.

Medical imaging technology, fueled by massive increases in computing power and new graphics algorithms, has made it possible for us to look more closely inside the body than ever before. Medical scans pass beyond the outer covering and look deep inside an organ to depict its hard and soft tissue, air spaces and tears and blockages. X-rays, the oldest and most widely used form of medical imaging, pass an electromagnetic beam through the body.

Ultrasound is another widely used imaging technique. Ultrasound works like a bat navigating its way through the dark by clicking out sound waves that bounce off of tissues and provide information about its shape and surface details. Another widely used imaging technology, medical resonance imaging, or MRI, magnetizes protons in the body's water molecules and transforms their vibrations into high-resolution images of the organs and soft tissues. Another imaging technique called positron emission tomography, or PET, detects gamma rays from radioactive material that's swallowed by or injected into the patient and is captured by a gamma ray cameras.

If you have ever watched a medical crime movie or viewed your own MRI or CT scan images, you probably noticed that a medical image isn't one picture, but instead, consists of dozens of images, each depicting a single cross section of a body part. Medical images such as MRIs and CT scans depict the body in hundreds of shades of grey. Bone appears white. Soft tissue resembles the subtle color variations of the sky on an overcast day. Voids, such as the insides of the lungs, appear as flat black.

In a medical image, different cell types that share a similar physical density appear in the image as nearly identical shades of gray. These confusing shades of gray are a major obstacle to turning image data into a 3D computer model that can be 3D printed. Somehow, the subtle shades of a medical image must be deciphered they can be modeled into a computer design file.

Gradations of white to gray to black don't provide enough information to guide a 3D printer through the process of making a complicated body part involving multiple cell types with a confusingly-similar appearance. Computers are getting increasingly better at identifying subtle patterns in even the slightest image gradations, and so there is plenty of hope. Turning grayscale images into crisp and meaningful digital data remains a major area of medical imaging research. Despite these limitations, however, medical researchers and surgeons are managing to 3D print an amazing variety of accurate and highly detailed artificial body parts and surgical models.

## The future

Thousands of people need new body parts if their original ones fail due to disease, a birth defect, or an accident. Fifty percent of people over 50 could use a replacement spinal disc. Despite the overwhelming demand, replacement body parts are hard to find, and they cost a lot. According to the United Network for Organ Sharing (UNOS), only 1 to 2 percent of the population manages to die in a way that makes them potential organ donors.[11] Even Steve Jobs, one of the richest men in the world, had to wait for his replacement pancreas, and he still died shortly thereafter.

If stem cells are the raw material of bioprinting, 3D printing complicated vascular systems remains the tissue engineering equivalent of the 4-minute mile. In 2004, researchers at the Medical University of South Carolina wrote that "assembly of vascularized 3D soft organs remains a big challenge."[12] Several years later, this still holds true.

As eloquently described by an article in *Science* magazine,

> *Without a vascular system—a highway for delivering nutrients and removing waste products—living cells on the inside of a 3D tissue structure quickly die. Thin tissues grown from a few layers of cells don't have this problem, as all of the cells have direct access to nutrients and oxygen. Bioengineers have therefore explored 3D printing as a way to prototype tissues containing large volumes of living cells.*[13]

Even if it were possible to 3D print living veins, however, that still wouldn't complete the job. Similar to the bioreactor-inflicted pounding required to process cartilage from petri dish into mature tissue, a highly vascularized body part—even if it were possible to print one—can't just be popped directly into the body—it needs to grow in. New arteries and veins must knit themselves into existing arteries and veins.

James Yoo, a professor at the Institute for Regenerative Medicine at Wake Forest University, described the challenge, "How can we create and connect those tissues produced outside the body? Whatever you put in the body has to be connected with the body's blood vessels, blood supply and oxygen. That's one of the challenges we'll face with larger tissues."

The biotech researchers of the future are already working on solutions. Yaser Shanjani, a postdoctoral fellow at Stanford University, will soon join the ranks medical researchers who view 3D printing and design software as critical tools in medical research. "I believe the future of tissue engineering will be incredibly integrated with 3D printing," he said. Yaser's specialty is printing bone grafts, or in more technical lingo "3D printing bioresorbable inorganic polymeric material," that after implantation eventually knit into the patient's body.

Yaser believes that a holistic approach that combines 3D printing technology, design software, and the body's own growth factors is the best path forward. Throughout his graduate career, Yaser has worked with interdisciplinary teams of biologists, surgeons, and material and manufacturing experts. "An ideal bone implant is an engineered construct that mimics natural tissue in terms of geometry, micro-structure and biomechanical behavior, and is eventually replaced with natural tissue," he explained.

I asked Yaser what he would do with the money if he were given several millions of dollars to spend on advancing bioprinting. His answer was swift and certain. "I would spend some of the funding on learning to print organs that are made of implantable materials seeded with bioagents (such as stem cells and growth factors). Next, I would like to see printers on site in operating rooms. Then, the big dream would to be send micro-robots into the patient's body where they could print new organs."

Printing living tissue and new organs directly inside a body would be a radical and lifesaving medical breakthrough. Printing inside the body could dramatically improve the survival rate of soldiers wounded on a battlefield or accident victims in emergency rooms. In less medically urgent situations, the ability to print precisely shaped living tissue would open up new areas of medical care and surgical training and enable new methods of drug development.

Artificial tissue or mini-organs that could be printed on demand would serve as extremely useful research test beds.[14] Artificial tissue could be used to study diseases, or to grow stem cells into mature, differentiated cells. Rather than testing out new drugs on mice and petri dishes—both crude approximations of humans—if artificial mini-organs could be 3D printed using a specific patient's cells, we could identify a drug candidate's effectiveness and side effects much more accurately.

3D printed artifical mini-organs could help young surgeons train. Cadavers have long been the training ground of choice, which poses several problems. Most training hospitals can offer their students only a random assortment of cadavers. In other words, people who've agreed to donate their body to science are frustratingly prone to dying of causes that don't map neatly to course curriculum or a particular research project.

A randomly selected cadaver works fine for teaching introductory level courses. However, for advanced students (and medical researchers) eager to delve deep into a particular specialty, this "one size fits all" method limits their options. For example, a medical school seeking to train students how to operate on a brain tumor faces a difficult, nearly impossible task in legally obtaining several fresh cadavers with brain tumors.

Until we can 3D print precise medical conditions, medical schools continue to do their best. Even in hospitals that have state of the art operating chambers, surgeons in training learn to operate using low-tech improvisations. When I visited a major teaching hospital, one of the teaching staff showed me how they trained students to do heart bypass surgery. The instructor stuffed two pieces of a tee-shirt into a closed box, and told me that students are asked to stitch the shirt's pieces together by inserting surgical tools through tiny holes punched in the side of the box.

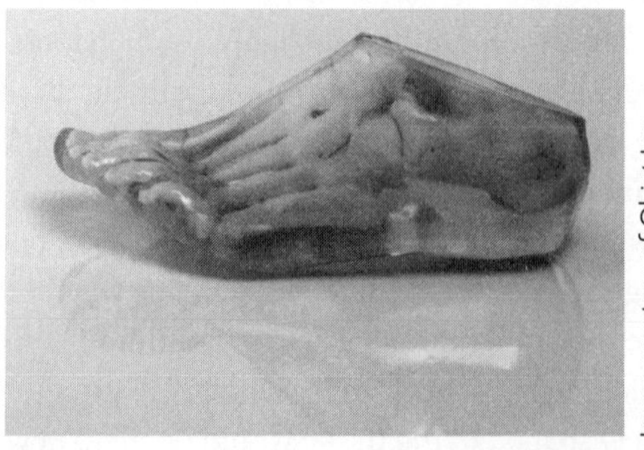

Image courtesy of Objet Inc.

**Printed model based on a CT scan for medical training**

3D printers could fabricate special surgical training models on demand. Printed training models could be carefully designed to mimic the properties of real tissue, organs, or even complete sections of the body. This way, until real bioprinting matures, medical schools could print out "dummy" parts that emulate the look and feel of the real thing, complete with fatal conditions and pathologies on demand.

Today 3D printed surgical models of bones or organs are already in use. Surgical planning, or surgical modeling, the dress rehearsal before a surgery where surgeons practice using realistic, true-to-size "practice parts" that represent the bones or organs they'll be operating on. To cut the time and potential mishaps of a real surgery, surgeons practice assembling, pushing on, even stapling together these practice parts. Surgical models also help surgeons communicate surgical procedures to the patient's families.

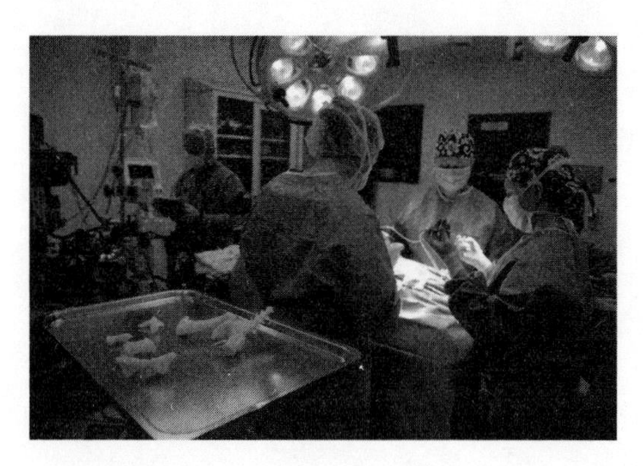

**Veterinarians practice an upcoming hip surgery for a dog using 3D printed surgical models of the dog's bones.**

3D printed surgical models and inanimate prosthetic body parts are just the beginning. Bioprinting will take personalized medicine to new heights. In the meantime, medical researchers and technologists face a broad array of barriers, from technological to biological to social to regulatory.

Our bodies are composed of thousands of different sorts of materials and today's 3D printers can print just a few materials at a time. Complex organs are full of blood vessels. Many critical organs such as the heart leave no room for technical glitches or adjustments. No one fully understands how to breathe the spark of life into artificial body parts. Even simple bodily organs function according to the precisely orchestrated interplay of thousands of different cell types.

Engineering tissue remains a delicate and difficult activity, bound up by ethical issues, political controversy, and (rightly so) stringent regulatory processes. Getting government regulators to approve new medical techniques can take years and millions of dollars in research. It may take years to encourage practicing surgeons, doctors and health insurance companies to accept bioprinting as a standard medical practice.

Rapid advances in medical and 3D printing technologies will transform medicine. Today's modern medicine would have looked miraculous if presented to someone living 100 years ago. Perhaps in 100 years bioprinting will be a commonplace medical process, the technological equivalent of the Fountain of Youth.

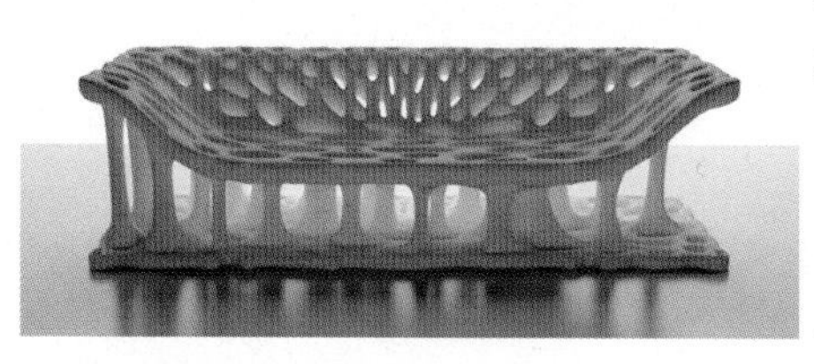

Image courtesy of Andrea Morgante and Enrico Dini, D-Shape

**A printed full-scale bench in stone-like material {Chapter 2}**

Image courtesy of Objet Inc.

**Multi-material 3D printing is in its early stages. This toy is actually a sophisticated engineering project made of several different raw materials that were blended together during the printing process. (Chapter 2)**

Image courtesy of Cornell University. Curator David I. Owen; Design: Natasha Gangjee; Photo by Jason Koski

**CT scanned priceless artifacts can be 3D printed for preservation and educational purposes. On the left is the original cuneiform and its 3D printed replica on the right. An enlarged image of the replica is below. (Chapter 2)**

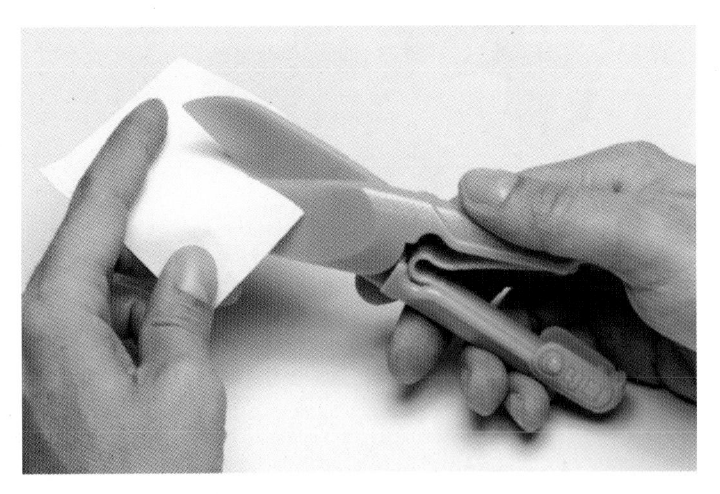

**Printing functional objects. These 3D-printed scissors work "out of the box"—no assembly or sharpening required. (Chapter 2)**

Image of vehicle printed on ZPrinter 650 courtesy of 3D Systems

**The plastic parts in this image look like assembled bricks but they were actually 3D printed, pre-assembled, in a single print job. (Chapter 2)**

Image courtesy of Kerrie Luft

**The titanium heel of this shoe was 3D printed in a single piece. (Chapter 2)**

Image courtesy of Olaf Diegel, New Zealand

**This printed electric guitar was made on a 3D Systems sPro 140 SLS system. (Chapter 3)**

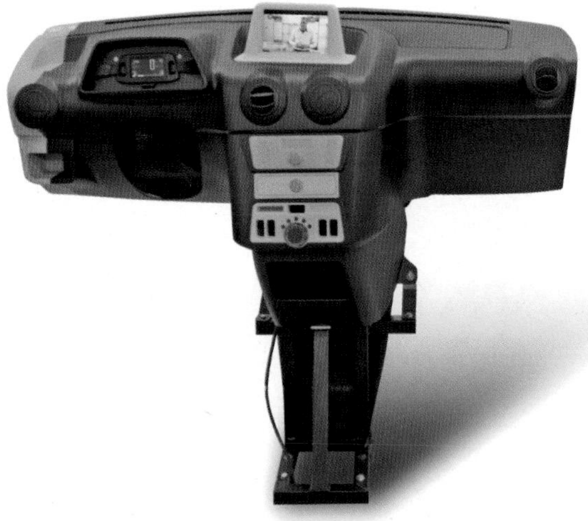

Image courtesy of Objet Inc.

**A 3D printed life-sized prototype of a truck cab, complete with working parts (Chapter 3)**

**The ultimate 3D printed architectural model: an ancient monastery (Chapter 5)**

**A PolyJet printer fabricating one of the candidate structures for this book's cover page (Chapter 5)**

Image courtesy of Solid Concepts Inc.

**As the UV laser traces the shape of successive cross sections, the solidified parts are slowly lowered into the tank. (Chapter 5)**

Illustration courtesy of Solid Concepts Inc.

**A laser beam melting and fusing powdered metal. The product will end up buried in the powder. (Chapter 5)**

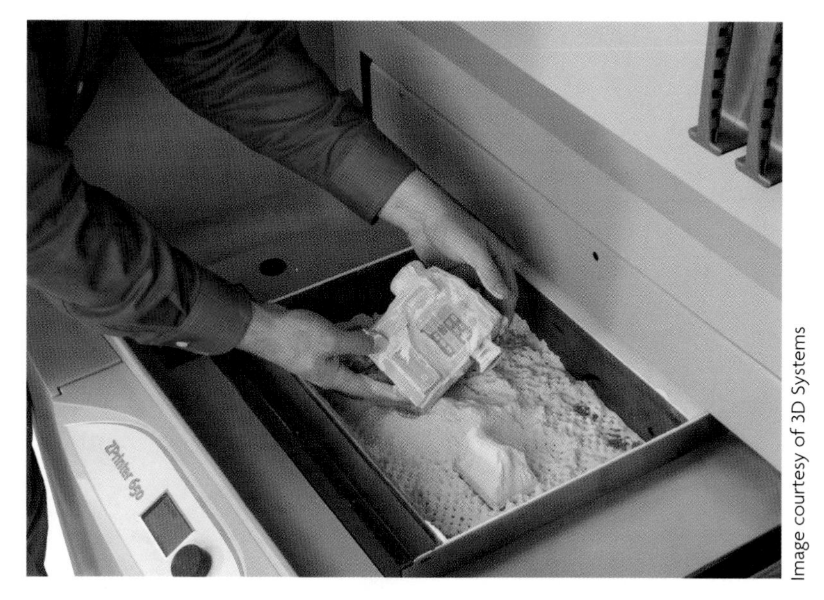


Image courtesy of 3D Systems

**The printed object is unburied from its powder print bed and cleaned up. (Chapter 5)**

Image courtesy of Gonzalo Martinez, Autodesk

**3D printout of a solid model of a turbojet engine modeled using Autodesk Inventor (Chapter 6)**

Image courtesy of 3D Systems

**A stylish artificial limb (Chapter 7)**

Image courtesy of Fraunhofer IFAM. Human lower jaw with biomimetically graduated cellular structure, made by SLM (metallic bone foam) © Fraunhofer IFAM.

**A 3D printed titanium jaw bone implant (Chapter 7)**

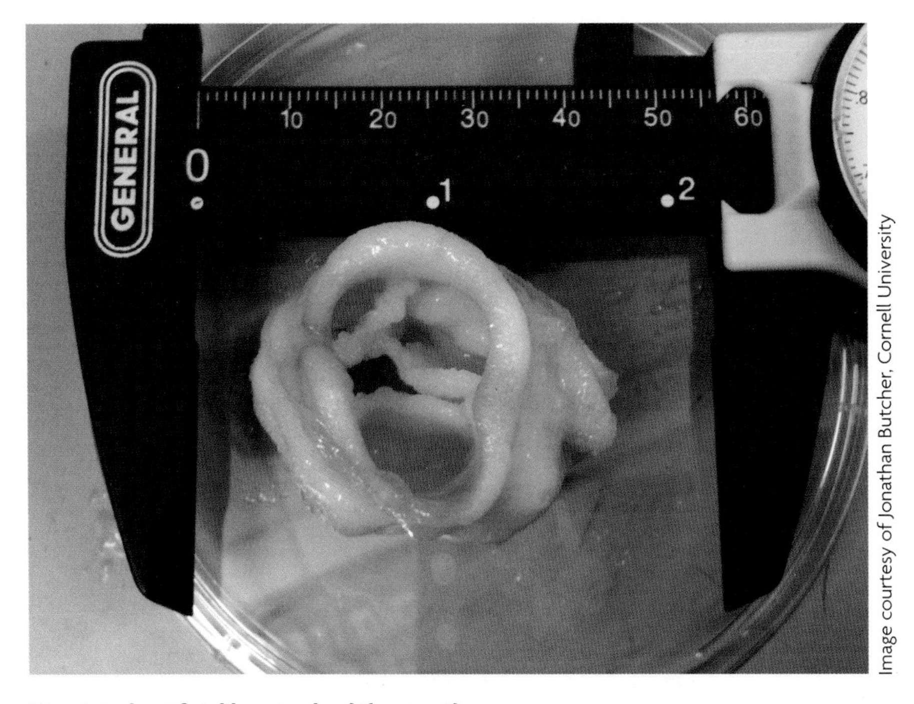


Image courtesy of Jonathan Butcher, Cornell University

**3D printed artificial heart valve (Chapter 7)**

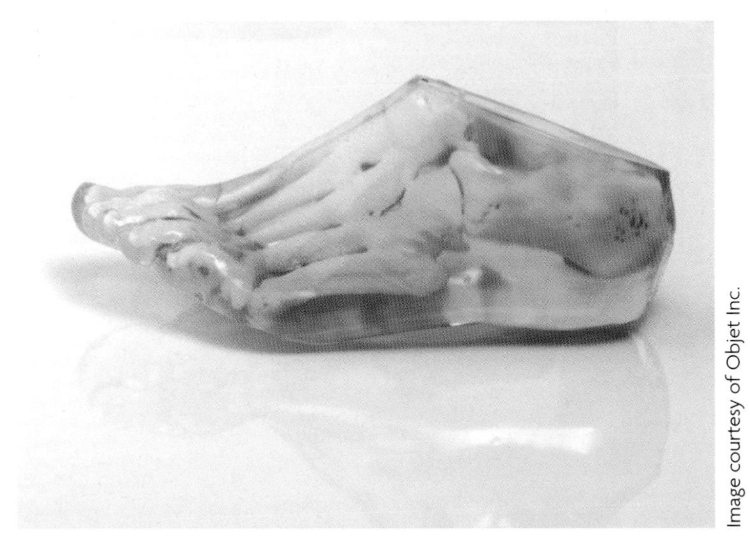

Image courtesy of Objet Inc.

**Printed model based on a CT scan for medical training (Chapter 7)**

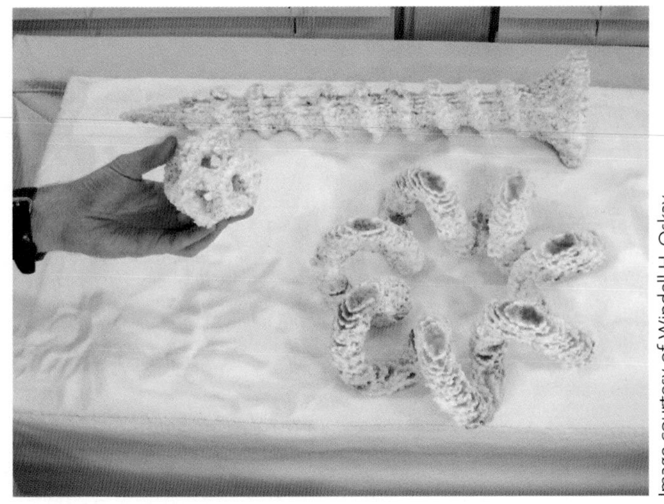

Image courtesy of Windell H. Oskay, www.evilmadscientist.com

**Heat gun caramelizes sugar to form 3D structures, no assembly required (Chapter 8)**

Image courtesy of Jeffrey Lipton, Cornell University

**3D printed cornbread in the shape of an octopus (Chapter 8)**

Image courtesy of Dave White, United Kingdom

**A 3D printed model of Mount St. Helens (Chapter 9)**

Image courtesy of MGX, a division of Materialise. Designed by Gernot Oberfell, Jan Wertel, and Matthias Bär.

**A coffee table whose shape is based on an algorithm that mimics the structure and growth of tree branches (Chapter 10)**

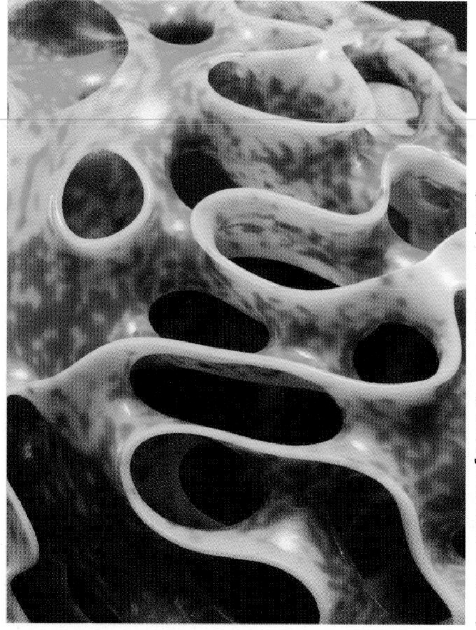

Image courtesy of Neri Oxman, W. Carter (MIT), Joe Hicklin (The Mathworks) Centre Pompidou, Paris, France, Photo: Yoram Reshef

**MIT professor Neri Oxman is one of the leading researchers in generative design. She uses 3D printing to fabricate shapes designed by mathematical algorithms to make forms that would not be possible to make using conventional materials such as glass, steel, and wood. (Chapter 10)**

**Oil Waterfall. Fluid simulation of oil was used to compute geometry, then "frozen" to create printed model (inset). (Chapter 10)**

Image courtesy of Ross Barber

**A 3D printed shoe designed by Ross Barber at the London College of Fashion and optimized by Within. The inner leather shoe was hand stitched in to make the shoe more comfortable when it was worn on the catwalk. (Chapter 10)**

Image courtesy of Jenna Fizel and Mary Huang, Continuum Fashion

**Next generation beach wear: a 3D printed bikini designed by Jenna Fizel and Mary Huang of Continuum Fashion (Chapter 10)**

Image courtesy of Enrico Dini and Andrea Morgante

**Enrico Dini prints structures out of a blend of sand and adhesives. The end result is a hard and smooth stone surface. The shapes and curves would be very difficult to hand carve out of stone or marble. (Chapter 10)**

Image courtesy of Sungwoo Lim and Richard Buswell
Photo: Agnese Sanvito

**The "Wonder Bench" is 3D printed out of concrete. The curved shapes and hollow interior channels would be impossible to make using traditional cement pouring techniques. (Chapter 10)**

Image courtesy of Michael Guslick

**A .22-caliber gun manufactured partly with 3D-printed plastics (Chapter 12)**

Image courtesy of Lee Griggs (Solid Angle SL) and Tomás Fernandez Serrano (forominecraft.com). Rendered with the Arnold global illumination renderer.

**A *Minecraft* scene created by designers Lee Griggs and Tomás Fernandez Serrano. If you look closely, you can see the Mario Brothers in red and green hats to the left of the central dome. (Chapter 13)**

Image courtesy of Nervous System, Inc.

**Generative designs can be used to design objects with a more organic look and feel. (Chapter 13)**

Image courtesy of Robert MacCurdy

**Concept demonstration of hybrid digital-analog print. Body printed in transparent analog (smooth) material. Inside is a visible digital lattice structure of mock-voxels. (Chapter 14)**

# 8 Digital cuisine

The new personal chef will be a 3D printer in your kitchen, one that's hooked up to the Internet to await text messages or email instructions about your next meal.

When I talk about 3D printing food, people's reaction is usually amusement blended with a tiny bit of revulsion. There's something about squeezing raw food stuffs through a print head that in many people's minds, is a bit, well, freakish. Interesting, certainly. But just too grotesquely processed.

Yet, food printing gets people excited. Once, outside a hotel in Washington, DC, the hotel's taxi stand manager saw me struggling to carry a boxy, unwieldy Fab@home printer. He abandoned his taxi stand and bounded over, but not to get me a taxi. He wanted a closer look at the 3D printer.

"I saw that thing on CNN!" the man shouted, pointing to the printer. "Are you the guy that was doing the food printing?" he asked. I confirmed that yes indeed, this machine and I were indeed the same duo that he had seen on TV. The man examined the Fab@home I was carrying and as he hailed the taxi, excitedly continued. "My wife and I watched that show, and she wants to get one of these. She's got a bunch of ideas for making Angry Bird cookies. And my brother-in-law – he should probably print himself some broccoli..."

Later that week at a conference of hotel executives, I gave a presentation about food printing. The conference's theme was "What Lies Ahead in the Hospitality Industry?" Throughout the day, a few dozen hotel executives listened and sat politely, occasionally scribbling down notes. To break the tedium, speakers and conference attendees periodically wandered to the back of the room to nibble on a gourmet snack or refill a coffee cup. Finally, after a long day of listening and eating my session on food printing began.

I put Fab@home on a table, loaded up with cookie dough, and started a print job. After a few minutes, the audience started craning their necks to see what the printer was doing. Chairs emptied. I paused, a bit confused and concerned as people started to migrate to the front of the room.

My audience dispersed and a crowd formed around the food printer. Dignity and senior rank forgotten for a few moments, executives clustered around the printer's plastic case, watching a nozzle zip back and forth extruding cookie dough into whimsical shapes. Upstaged by an automated cookie maker, I hastily wrapped up the presentation and turned to the question and answer session.

People returned to their seats and started brainstorming. A marketing executive suggested that their loyalty program should design custom food and offer fresh, 3D printed goodies to guests at the check-in desk. Another executive suggested that her hotel could offer themed 3D printed snacks on their room service menus, or could fabricate custom snacks to appeal to each guests' unique pallet or dietary needs.

My experience has been that everyone—no matter their profession or rank—is interested in food preparation. 3D printing food is still in its infancy, the domain of a few bold gastronomical adventurers and academic researchers. Yet, like microwaves and automated coffee makers, food printers have tremendous social appeal.

## Digital gastronomy

You won't see a commercial 3D food printer at your local electronics store yet. But if we fast forward a few years, perhaps we'll see food printers in home appliance stores that look something like The Cornucopia, or the "automated horn of plenty,"[1] a design concept for a family of four food printer prototypes: the Digital Chocolatier, the Digital Fabricator, the Robotic Chef and the Virtuoso Mixer.

The design team that created the Cornucopia concept, led by Marcelo Coelho and Jamie Zigelbaum, launched the design concept on their website by posting a series of stunning images of each futuristic prototype. After the design images of the Cornucopia family of food printers went live, they looked so realistic the pictures created an immediate storm of interest on the Internet. People weren't sure whether the design concepts pictured were real food printers or just skillfully rendered design concepts. Eventually, the confusion was sorted out and (much to the disappointment of 3D printing enthusiasts and technology-inclined foodies) food fans learned that the Cornucopia prototypes were not yet commercially available products.

The intent of the Cornucopia design concept is to transform cooking. The project's website explains, "While digital media has transformed every facet of society, the fundamental technologies we encounter in the kitchen today provide only incremental improvements to the tools we have been using for hundreds of years." Each Cornucopia prototype addresses a fundamental cooking process. What makes them disruptive is the application of computing to foodstuffs. On their website Coelho and Zigelbaum explain that their design prototypes could offer a new way to cook by physically and chemically transforming "ingredients into new compounds; and finally their modeling into aesthetically pleasing and delectable textures and shapes.

Illustration courtesy of Marcelo Coelho

**The Digital Chocolatier is a design concept for a food printer that can create custom chocolates.**

The strong emotional and aesthetic appeal of the Cornucopia food printer prototypes demonstrates the potential for technological innovation in cooking and eating. Food printers would offer new creative opportunities. Cookie cutters and cake molds, like traditional manufacturing machines, have their limitations when it comes to forming precise and complicated shapes and patterns. In contrast, digital gastronomy, according to Zigelbaum and Coelho, would allow "for the creation of flavors and textures that would be completely unimaginable through other cooking techniques."

If they were real, each food printer would feature a touchscreen interface, a built-in memory card to allow users to save favorite recipes, and web connectivity so users could control the origin, quality, nutritional value and taste of every meal.[2] The appealingly named "Digital Chocolatier" prototype would enable users to quickly design, assemble and taste different chocolate candies. According to the design, the Chocolatier's ingredients would be stored in metal cartridges placed on a carousel. If I could buy a Digital Chocolatier, I would want one with a "publish" button so I could post a particularly successful chocolate creation onto a website or send it to a friend.

The second prototype of the Cornucopia design concept is the "Digital Fabricator," a personal, three-dimensional printer for food. As described on the Cornucopia website, the Fabricator's cooking process would start "with an array of food canisters which refrigerate and store a user's favorite ingredients." The Fabricator printer concept would store, mix, deposit and cook layers of ingredients, allowing for the creation of flavors and textures that would be unimaginable through conventional cooking techniques.

The images for the Digital Fabricator show a gleaming machine the size of a microwave oven. A gleaming rack of silver food cartridges, each tucked into a metal coil, have the sleek appeal of a collection of Williams-Sonoma cocktail shakers. A computer would oversee the Fabricator's food preparation process. Each food cartridge would be instructed to pipe its raw ingredients into a central mixer where they would be blended and pushed through an extruder head into accurately deposited elaborate food combinations.

The final two members of the Cornucopia family of futuristic food printers are the Robotic Chef and the Virtuoso Mixer. The Robotic Chef concept would physically and chemically transform a single solid food object, such as a steak, fish or a fruit. The Virtuoso Mixer concept depicts a machine composed of a three-layer rotating carousel that could provide cooks with an efficient way to mix multiple ingredient variations and experiment with subtle differences in taste and composition.

I hope that someday kitchen food printers will be as simple to operate and as appealing as the Cornucopia concepts. Food printing, or digital gastronomy, would offer new creative culinary frontiers and also a convenient mode of food preparation for busy people. Food printers, digital recipes and corresponding food cartridges would enable cooks to create novel and nutritionally balanced delicacies, opening up new culinary markets.

Illustration courtesy of Marcelo Coelho and Amit Zoran

**Concept designs for digital gastronomy fabricators**

Unfortunately, digital gastronomy remains mostly conceptual. Big food manufacturers and companies that sell household appliances haven't yet launched commercial-grade food printer products. As a result, most food printing today takes place in research labs, where students, scientists and engineers find their way to printing food by accident while trying to solve an unrelated engineering or design challenge.

Food pastes are a good test material for scientists and engineers. Food is cheap, plentiful, and non-toxic. Given the diversity of available raw ingredients, a resourceful investigator can find food ingredients that mimic the material properties of much more expensive and rare printing materials.

My students and I accidentally stumbled into food printing as an engineering accident with the Fab@home printer. It started when one of my graduate students realized that cake frosting makes a great raw material for prototyping engineering design projects and calibrating printer settings. Cake frosting dissolves in water so it's easy to rinse off in the sink . . . or perhaps lick off when no one is watching. Cake frosting quickly became the most commonly used sacrificial material for 3D printing projects.

My students and I weren't the only people venturing into 3D food printing. Several years ago when we launched Fab@home into the world, we expected that people would build printers and use them to print plastic parts or perhaps toys or useful things for their homes. Instead, Fab@home users wrote to us to share stories of printing food. Noy Schaal, a high school student in Louisville, Kentucky, was one of the earliest users of Fab@home when we open sourced the machine designs in 2006. With her father Maor, Noy built her own Fab@home printer and then customized it, adding a heated chocolate extruder.

**The printed chocolate that won high school student Noy Schaal first prize at a high school science fair in 2006**

For several weeks the Schaals experimented with different temperatures and printing nozzles. Their efforts paid off at the state's high school science fair where Noy designed and printed chocolate pieces shaped like the state of Kentucky, securing first prize.

## Like a computer-guided frosting bag

Printing even simple food paste is a thorough workout in applied food engineering. Like many of life's pastimes, 3D printing food is a lot harder than it looks. Printed food must be made with just the right application of mechanical force, plus a well-designed digital "recipe" using raw ingredients of the correct consistency. Raw ingredients must be soft enough to push through a syringe tip, yet stiff enough to hold a shape after being printed. Add to that the challenge of dealing with each foodstuff's unique material properties, different temperature tolerances and different cooking methods.

A hamburger isn't complicated to make if you're doing it the old fashioned way by grilling meat, slicing a tomato and onion, squirting on ketchup and then placing the whole creation into a bun. On a 3D printer, however, a simple burger becomes a complicated, multi-material food engineering challenge. 3D printing a delicious fresh hamburger with all the right condiments, even if it were possible, would be a pretty serious feat of culinary engineering.

Printing out raw ground meat in the shape of a patty is simple. Printing some ketchup on top wouldn't be a problem either. Even printing and baking the raw bread dough for the bun would be mostly a matter of determined engineering.

Food printing becomes difficult when making foods that are fresh, or made by nature. Printing fresh and delicious slices of tomato, onion and lettuce that taste good and look "real" launches you into the world of hard-core, industrial-level food engineering. 3D printing a fresh, hot hamburger with everything on it would be a feat of engineering as difficult as 3D printing a complex organ. That's why academic researchers and technophile foodies must still stick to designing and printing simple, soft foods.

The basic food printing process goes as follows: like any other 3D printed goodie, food concoctions start as a design file. However there's no commercial software that could be considered "Food CAD" yet. Food printing aficionados download their food design file from a website or create their own design files from scratch using engineering design software.

Consumer-level printers such as the RepRap or MakerBot's popular Replicator printers were built to print spaghetti strands of plastic from reels that feed into the print head. To print food, the printer must first be specially outfitted so it can work with edible raw material. 3D printers whose blueprints are open sourced are ideal for the customization that food printing currently demands.

Most people transform their plastic printer into a food printer by attaching syringes full of food material to the print arm. Some printers use a motorized plunger that applies pressure to the food syringes. Keeping the motor's pressure at the right point involves calibrating the motorized plunger for each print job to apply exactly the right degree of force. A human cook instinctively knows when to apply pressure and when to use a light touch. In contrast, a food printer doesn't have that knowledge.

One leading commercial consumer printer, MakerBot's Replicator, can be transformed into a food printer if it is fitted up with an add-on part called a "Frostruder." The Frostruder is clamped to the inside of the 3D printer. Plastic clips hold the printing nozzle firmly in place and users can reload fresh food into the nozzle using a twist-off cap.

The nozzle pushes out the food material using pressure made by an attached air compressor. A Frostruder is capable of handling up to 100 pounds of pressure per square inch—enough to pump up a bicycle tire. However, the company's website warns users to "never exceed 100 PSI." They don't specify exactly what would happen, but I can imagine a small explosion that would spray the kitchen walls with peanut butter, Nutella, or some other similarly sticky goo.

Different recipes call for different degrees of mechanical force. Sometimes room temperature affects the flow of food paste through the nozzle. The size, or diameter, of the printing nozzle is critical. If the nozzle is too small, the food paste won't come out fast enough. If the nozzle is too big, the printed goodie will appear crude and rough.

Even if the printing nozzle is the right shape and the food is flowing at the correct rate, some raw food stuffs just don't behave themselves in the syringe. Sometimes oil will pull together in clumps. Or water in a food paste will sink to the bottom of the syringe, resulting in a misshapen piece of printed food that looks nothing like its creator intended.

One of the most confusing aspects of 3D printing food is the way it's cooked. People frequently misunderstand the process and think that the print nozzle will somehow deposit a ready-made fried chicken breast or baked bread. However, today's 3D food printers don't have the technical capacity to sauté, grill, or fry what they print. It is possible to bake cookies, however. A printer "bakes" cookies by using a heated build platform that sits under the print head. As each cookie is printed out, the heated platform bakes the raw cookie dough.

## High res and low res . . . shortbread cookies?

3D printed food that holds its shape and looks attractive is all about resolution. In the old days, when people used the words "resolution" and "food" in the same breath, they were usually referring to the struggle to stick to their diet. In the brave new world of 3D printing food, when experts talk about resolution, they're talking about how well a printed food holds its shape. According to Brandon Bowman, a former journeyman blacksmith and now a student in the Solheim Additive Manufacturing Lab at the University of Washington, shortbread cookie dough is an ideal raw material for 3D printing. When shortbread dough is baked, it "can hold its resolution enough to print out the individual teeth on a gear," said Brandon. Unlike chocolate chip or peanut butter cookie dough, shortbread cookie dough can endure high pressure and not fall apart in the oven.

Image courtesy of Jason Bowman, University of Washington

**A delicious 3D printed, high res shortbread cookie**

Brandon found his way to food printing while working on a research project in tissue engineering. His research involved applying 3D printing technologies to help victims of severe burns quickly re-grow destroyed bodily tissue. Brandon's initial research goal had been to print biodegradable tissue scaffolding whose shape would speed up the growth of new skin and then dissolve, leaving the new skin to grow in peace.

Since bioprinting artificial tissue is a cutting edge field, researchers create their own printing materials. Brandon's material of choice was ground up crab shells mixed with other ingredients to give it the right consistency. "I spent winter quarter catching crab cells, drying them, and grinding them into powder to make a paste," said Brandon. Catching crabs in the icy waters of Puget Sound, particularly during Seattle's long dark winters, is slow and chilly work, however.

Brandon found that he was rapidly wasting precious crab powder due to mechanical problems with the printer and the texture of the crab paste. After the paste was printed, it would separate or fail to hold its shape once it left the nozzle. As his valuable mixture was rapidly used up, Brandon realized that maybe there was another way to experiment with the paste's material properties.

He needed a test material that would be similar in texture and behavior to ground up crab shells. The answer turned out to be a blend of salt, flour, sugar, butter and water. Or shortbread cookie dough. "I decided that if I can make shortbread cookie dough that will hold its shape, that would help me make tissue scaffolding to help burn victims grow back their skin," Brandon explained when I spoke to him on the phone.

"The nice thing about cookie dough is that it's cheap and easy to get," said Brandon. "My friends who are bakers thought it was an odd way to do research but for me, it was a nice change to end up with shortbread cookies to eat at the end of a long day in the lab," he laughed. Brandon patched together a custom-made food printer—a RepRap Prusa fitted out with a MakerBot Frostruder—plus some Luer locks, or nozzles used in medical devices.

Here's Brandon's recipe for 3D printed high-res shortbread:

1 cup of flour
½ cup of powdered sugar
Mix thoroughly
1 stick of butter
½ cup of honey
1 teaspoon of vanilla
¼–½ teaspoon of salt (optional)
Makes 10-15 50cc syringes print with Green Luer lock.
Bake 7-12 minutes at 350 degrees F

A printer "bakes" cookies by using a heated build platform that sits under the print head. As each cookie is printed out, the heated platform bakes the raw cookie dough. I like to call this type of baking "inline cooking."

Even a simple shortbread cookie recipe is a complex engineering process. Brandon described some of the challenges he faced. "Any baking soda, or baking powder causes rapid expansion. Water in the recipe will cause the printed cookie to slump down and spread out." And of course there's the aesthetic component. "Finding the right balance between ease of printing and baking resolution while maintaining taste was the hardest part," said Brandon.

## Cookies with text inside

One of the wonderful things about eating a piece of lasagna is its internal composition, the rich variety of different types of cheeses, mushrooms, and pasta in the gooey inside and the rough texture of the hardened baked pasta on top. However, multi-material printing is still in its infancy.

A few years ago Franz Nigl, then at Cornell, printed multi-material cookies of two different colors of cookie dough. Franz was already a skilled baker. He was from Austria, a country with a proud tradition of precision baking. In fact, his grandmother was known for her fine desserts, a standout baker even in Austria's crowded field of gourmet pastry chefs.

Franz's goal was to prove that multi-material 3D printing could be applied to fabricate objects with a complex internal geometry. Franz's design project was to print cookies whose insides would contain a special surprise for their "user." Eventually, like Brandon, he found that cookie dough was ideal raw material.

Image courtesy of Jeffrey Lipton and Franz Nigl

**3D printed cookies with the letter "C" inside**

The surprise would be an internal pattern that would appear unexpectedly when the cookie was bitten into. Franz created a design file and set up an experimental test platform on Fab@home using two different print nozzles, one with chocolate cookie dough and the other containing vanilla.

The first batch of printed chocolate and vanilla cookies looked great when they were squeezed out of the printer's nozzle. The CAD file worked. The problem was that when baked, the raw cookie dough quickly lost its shape in the oven, melting into shapeless lumps. Franz's printed cookies tasted good but looked like greyish lumps of coal. Worse, the surprise design on the inside was a mess, a blurry smudge of white and brown cookie doughs.

After some soul searching and a bit of additional research into whether cocoa or vanilla might have previously undiscovered material properties that behave in an unexpected manner when forced through a nozzle, Franz was ready to try again. He remembered that when he was a child, his grandmother used to make a cookie known for its beautifully crisp physical shape. He called his family in Austria and after a bit of explanation, was able to obtain the closely held recipe for his grandmother's cookies.

With fresh hope and a new batch of vanilla and chocolate cookie dough, Franz loaded up the printer. The CAD file and Fab@home took care of their part of the experiment, depositing perfectly formed cookie dough onto the build platform. Then, the big test: whether this new batch of cookies would hold their shape, both inner and outer, in the oven.

After an anxious 28 minutes, the cookies were pulled out. Like the first batch, they smelled delicious. Unlike the first batch, their outer shape was perfect. The old family recipe had come through.

The moment of truth lay in the next step: would the surprise design on the inside also hold its shape? Once the cookies cooled, Franz took the first anxious bite and held the cookie aloft for all to see. Triumph! Inside the cookie, as defined by project specs, the complex internal geometry, as revealed by the first bite, was a perfectly defined chocolate letter "C."

The potential commercial market for cookies with custom printed insides could be vast. Printed cookies could become vehicles to convey confidential information back and forth. Perhaps printing a new password into the inside of a fresh batch of cookies could help IT departments sweeten the annoyance of customers over headquarters-mandated computer password changes.

## Evil Mad Scientists print a gigantic sugary mathematical formula

Not all low-cost food printers use a nozzle. Lenore Edman and Windell Oskay, the founders of a relentlessly innovative small company Evil Mad Scientist, invented CandyFab, a 3D printer that uses a heat gun to melt raw sugar into rock-hard intricate shapes. Their design goal was to create a low-cost food printer that could use low-cost, recyclable material.

Their solution was inspired by industrial-scale 3D printers that use a laser or light source to firm up powered polymer or metal. CandyFab uses a heat gun to melt sugar, or what its inventors call "selective hot air sintering and melting" to fuse raw sugar. CandyFab creations have a rough, stone-like appearance, or what a technologist might describe as a low-resolution shape. The printer uses a blend of open source and commercial software.

Edman and Oskay created the CandyFab out of common household objects. The heat gun to melt sugar was a $10 air heating element, what Windell described online as the "baby sister of the one in your hair dryer." They attached the heating element to a cooling fan to control the temperature of the sugar during the printing process. Next, the Evil Mad Scientists attached the entire contraption onto a mechanical system recycled from two old HP plotters. They created the body of the CandyFab from wooden boxes covered with heavy canvas stitched together on a home sewing machine.

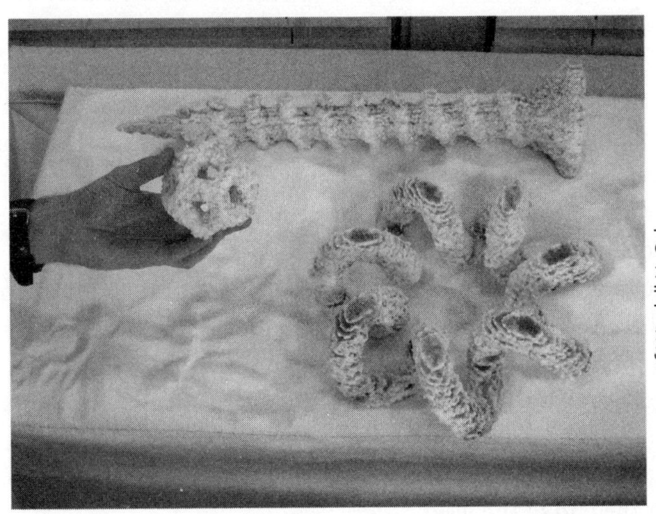

Image courtesy of Windell H. Oskay, www.evilmadscientist.com

**Heat gun caramelizes sugar to form 3D structures, no assembly required**

One of my favorite parts of the story is CandyFab's testing process. To test the precision and movement of the heat gun, Edman and Oskay placed a piece of bread under the gun, on the print bed. Its inventors knew they had succeeded when the CandyFab's heat gun slowly toasted the phrase used by worldwide by software developers when they create a new application, "Hello World."

## Feeding the quantified self

Precision food printers are the ideal output device for a era where diet, health and medicine are increasingly driven by data. Low-cost sensors, online assessment tools, low-cost DNA testing, and greatly improving medical diagnostic tests have paved the way for a new movement in bodily awareness, the Quantified Self or "self-knowledge through numbers." 3D printed food cartridges could be a food delivery vehicle for the growing numbers of people who count, log and analyze every biometric they capture.

Our bodies are the target of increasingly more sophisticated data collection. Quantified Self enthusiasts strap monitoring devices to themselves. They track how far they walk, their heartbeat, weight, calories burned, and how well they have slept. Imagine this emerging world of biometric data, combined with personalized medical care, combined with a home-scale, digitally controlled food printer.

It's easy to imagine a medical 3D printer fabricating the future version of granola or pharmaceutically-enhanced candy bars. The printer could be even more responsive than that, however. A kitchen 3D printer could be updated by the hour on its owner's medical issues. It could use its digital intelligence to mix and blend raw ingredients to print out meals specially tailored for clients' daily needs.

For most of human history, monitoring one's biometrics has been a fairly primitive process. People count their pulse, their breath rate, examine the surface of the tongue or the appearance and smell of their bodily waste. Now new medical technologies enable people outside the medical establishment to track and monitor, even predict, what's going on in their bodies. Other fields have been transformed by growing amounts of available data, computing power and the Internet. People can predict and manage non-intuitive causalities between, say, their sleep patterns the week before and a slide into depression at the end of the week.

The first time I heard about the "Quantified Self" movement, I was having breakfast with an old friend of mine who lives in the San Francisco Bay area. A third person joined us—her friend, a woman who was building a startup that involved a web app that enabled people to upload and analyze data about their bodies. As I sat happily sipping coffee and eating bacon, it took me a while to notice that my West Coast friends were breakfasting on decaf green tea and egg white omelets.

Our discussion centered on a website that was an online gathering place where people recorded, logged and talked about their biometric data. Site users recorded their pulse rate, what they ate that day and their exercise levels. Some users uploaded blood sugar readings. Others did psychological self-assessments online or logged their medications.

During the discussion, I realized that a food printer would be the perfect food technology in a digital era. Diabetics could upload blood sugar data, software could calculate the nutritional balance of their next meal and send a recipe to their kitchen 3D food printer. Health conscious users could print their morning toast with pre-selected nutrients already embedded.

Data-driven food printing would enable people to adjust recipes for a particular blend of nutrients in a particular portion size. Biometric data, combined with a networked 3D printer, could help people moderate their food intake to reflect their activity levels of that particular day. Food printers could help our overweight society move away from mass-produced processed food with limited variety and unlimited shelf life towards freshly printed edible creations.

A dedicated 3D printer chef could be a stern disciplinarian. A couch potato who skipped his morning jog would be denied his request for two pieces of printed pizza. After reading biometrics, his printer would instead print him a fresh Caesar salad and piece of whole wheat bread.

Software tycoon and mathematical genius Stephen Wolfram is a dedicated, long-time collector of personal data. On his blog on his company's website, he wrote, "[B]ecause I've been interested in data for a very long time, I started doing this long ago. I actually assumed lots of other people were doing it too, but apparently they were not. And so now I have what is probably one of the world's largest collections of personal data." [3]

Wolfram meticulously logs his daily habits. For decades now, he has kept track of the number of computer keystrokes he hit in a day, how many steps he walked, the number of hours he spent sleeping and on the phone. He wrote, "Every day—in an effort at 'self awareness'—I have automated systems send me a few emails about the day before."

Wolfram is a data analytics pioneer. However, as people become increasingly fascinated with analyzing their personal data, soon Wolfram's habits will be the norm. The Quantified Self movement is growing rapidly. In a way, it's about technology—DNA analysis, sensors, and analytical tools. But it's also about a person's relationship with their body.

Biometric data and precise, multi-material computer-guided food production will open up new frontiers in health. A 3D printer, complete with cartridges of foodstuffs optimized precisely for your bodily needs, will read wireless signals in real time from sensors on your body. Like a personalized chef and nutritionist in one, your kitchen 3D printer will print out your perfect meal, timed exactly to the minute you walk through the front door. Your printer will read GPS data from your car so it will know if you're caught in traffic or lingering late at work.

## Processed food

Printing custom food, even healthy and nutritionally optimized food, raises philosophical questions and stirs people's emotions. 3D printed food is processed food. Like bioprinting, tissue engineering, and particle accelerators, food printing could be viewed as a direct assault on the natural world.

Processed food is frequently blamed (and rightfully so) for contributing to diseases of modern civilization: obesity, cancer, and heart disease. In developing nations, critics of processed food point out that mass produced processed food imported from faraway places has cannibalized traditional, more eco-friendly and nutritionally valuable methods of food production. Trans fats, refined grains, excess salt and high fructose corn syrup are villains when it comes to maintaining a lean, well-nourished body.

Mass-produced processed food, sometimes called "Big Food," is prone to massive outbreaks of bacterial infections. Fast food chains have sickened hundreds of customers by serving infected meat. A typical processed food contains dozens of synthetic ingredients that give food its texture and color, preserve it, sweeten it, or somehow enhance its flavor.

Worst of all, processed food can taste bad. True, potato chips taste pretty good. But there's no comparison between a mass produced grocery store peach and a peach (fresh from the tree) bought at a roadside stand in New Jersey in July.

No wonder people frequently shudder or laugh when they learn about 3D printed food. An article in the *Reader's Digest*, a widely read popular magazine in the United States, eloquently expressed the downside of processed food:

> *Unfortunately, most processed foods are laden with sweeteners, salts, artificial flavors, factory-created fats, colorings, chemicals that alter texture, and preservatives. But the trouble is not just what's been added, but what's been taken away. Processed foods are often stripped of nutrients designed by nature to protect your heart, such as soluble fiber, antioxidants, and "good" fats. Combine that with additives, and you have a recipe for disaster.* [4]

The perceived evils of processed food are nothing new. In Battle Creek, Michigan, in the early 20th century, people journeyed to the Battle Creek Sanitarium to solve their health problems. The Sanitarium was founded by the Kellogg brothers who would ironically, would later go on to become wealthy by selling Kellogg's breakfast cereal, one of the most processed foods there is.

The Battle Creek Sanitarium offered what at that time was a radical new approach to health. Patients ate low-fat, low-protein diets and took part in vigorous exercise. The Kellogg brothers advised their wards "if you want to cure what ails you, eat like a chimp."

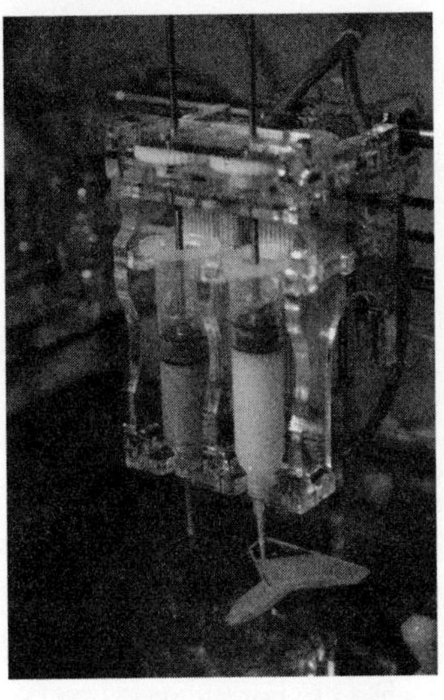

Image courtesy of Daniel Cohen and Chef Dave Arnold, French Culinary Institute, New York

**Printing with celery and scallops**

Actually, in terms of diet, the Kellogg brothers were on the right track about eating like a chimp. However, unlike chimps, humans can't easily live off of raw food gathered from their immediate environment. Chimps spend several hours each day chewing what they gather to obtain the nutrients they need.

Contrary to what many people believe, processed food is a major reason why modern humans lead long and healthy lives and enjoy leisure time. In an article in the *Utne Reader*, author Rachel Laudan, author of the book *In Praise of Fast Food*, questions what she terms "culinary Luddism." Industrialized food has brought about tremendous improvements to human welfare. Techniques to process, preserve and transport food have reduced malnutrition and given people time to pursue other activities.

Laudan writes that the "sunlit past of the culinary Luddites never existed. So their ethos is not based on history but a fairy tale."[5] Food technologies offer liberation and health. The freezer, a once a disruptive new technology, keeps food fresh for much longer. Agricultural fertilizers make food more abundant.

Laudan describes the often overlooked benefits of processed food. "[C]ulinary modernism had proved what was wanted: food that was processed, preservable, industrial, novel, and fast, the food of the elite at a price everyone could afford. Where modern food became available, people grew taller and stronger and lived longer."[6]

Image courtesy of Daniel Cohen and Chef Dave Arnold, French Culinary Institute, New York

**3D printed and fried scallops**

Preserving techniques such as pickling, curing, canning and vacuum packaging create processed food that's nutritious and can be transported and stored. Processed food brings food to people who live far away from its source. Processed food is less prone to rot. It can travel further from its origins. It brings regular people delicacies once reserved for aristocrats.

Digital cuisine, or food printing, like other life-giving food technologies, will introduce new health and social benefits. Biometric data and computing power will enable new combinations of ingredients and new food shapes. Food printing will introduce a new generation of processed food that's nutritious, cheap, fresh and delicious.

## From processed to synthetic food

If processed is controversial, how about food that's completely synthetic? I'm using the term "synthetic food" fairly loosely, meaning food that's edible, nutritious, even tasty, but did not originate from base ingredients that most of us would consider natural, or even recognize.

There are two ways to 3D print synthetic food. The first relatively straightforward method would be to mix up a paste of familiar food, say garlic scallop paste, and then to 3D print that into a novel shape. The second, more futuristic method of printing food would be to mix up a paste of chemical building blocks and use a digital recipe to 3D print raw food material into imitations of "real" food.

If you've ever taken apart a laser printer, you probably remember that only three core colors of ink—cyan, magenta, and yellow—can create a rainbow of different sorts of colors. By blending just a few basic colors together in exquisitely precise ratios, a laser printer can create a colorful and visually appealing document. When the concept underlying color printing is applied to the world of food ingredients, a new limitless world of culinary possibility opens up. Using just a few basic ingredients, a food printer of the future could combine raw food materials in new and different permutations to create an infinite number of new and diverse food types.

Image courtesy of Cohen, Lipton, Cutler, Coulter, Vesco, Cornell University

**Basic food components lined up for a food-printing trial**

Someday we will print synthetic nutrient building blocks to resemble familiar favorites such as grilled salmon with mashed potatoes and broccoli. We will print synthetic fish and beef to conserve the environment from overfishing and the high environmental cost of raising beef. Printable synthetic meals could nourish soldiers in difficult conditions. Refugee communities could have access to printed food that wouldn't spoil and would be relatively easy to carry.

Our planet is struggling to cope with population growth and an increasing demand for meat. Two major food sources are extremely abundant on the earth: insects and algae. In many countries people roast grasshoppers and larvae. Yet, Westerners aren't ready to eat insects.

Two entomology professors at a Dutch University, Marcel Dicke and Arnold Van Huis, are promoting insects as food. In an article, in *The Wall Street Journal,* they extolled the virtues of eating a "six legged meal."

> *Insects are high in protein, B vitamins and minerals like iron and zinc, and they're low in fat. Insects are easier to raise than livestock, and they produce less waste. Insects are abundant. Of all the known animal species, 80% walk on six legs; over 1,000 edible species have been identified. And the taste? It's often described as "nutty."*[7]

A few restaurants in the Netherlands have placed insects onto their menus (usually locusts and beetle larvae) but they haven't yet become major attractions. People may never willingly eat a whole freshly toasted beetle. But if you ground up insect parts and liberally sprinkled them into different colored gels, they could be printed into the shape of a delicious meatball.

Dan Cohen, while he was still a graduate student, as part of his graduate thesis, explored the production of truly synthetic food, or what he described as "a bottoms up approach to designing food." Dan is an engineer by training, not a chef. To begin his venture into 3D printing synthetic food, his first stop was across the street at Cornell's renowned School of Hotel Administration and Hospitality to recruit students well-versed in fine cuisine for the project team.

Dan's goal was to create many diverse types of food from as few raw ingredients as possible. This was a problem of optimization, of forming the largest number of combinations from a small set of building blocks. In contrast, the hotel school students had a different angle. They were intrigued with the challenge as it related to a course on artificial food enhancers they were enrolled in.

Dan and his team of hotel school students set about systematically combining base food ingredients in as many combinations as they could. They printed different combinations of flavorings, colors, artificial fibers, nutritional supplements, and texturing agents. They blended in various amounts of different gels and gums. Like the Oompa Loompas in Willy Wonka's Chocolate Factory, students ran food printing experiments day and night. Every day Dan would present to me new types of synthetic foodstuffs that resembled milk but were formed in the shape of a soft cube, or mushrooms that were brown but tasted like bananas.

The project turned out to be a resounding success from the perspective of food engineering. However, it was an equally resounding flop as a culinary technique. The printed synthetic foods were edible and didn't even taste that bad. However, they were just too weird. Nobody, not even the team of students who took part in the project, wanted to eat them, much less order them in a restaurant or buy them from a grocery store.

Image courtesy of Jeffrey Lipton, Cornell University

**3D printed cornbread in the shape of an octopus**

Dan's food engineering project proved that food manufacturing is both an art and a science. Consumers are unpredictable in their acceptance of commercial synthetic food. The key is that a synthetic food be recognizable. Candy that's strawberry flavored must be red. Artificial chicken meat made of synthetic protein must look like chicken, feel like chicken and taste like chicken. When 3D printed synthetic food products eventually appear in the culinary marketplace, they will need to look like a nice plate of sushi rolls or a baked duck.

Printing synthetic food that looks "real" is difficult to do. Printing fake fruits and vegetables is even harder. The challenges of printing food that's fresh or natural are similar to those faced by medical researchers trying to print living tissue. Nature-made foods are made of a complicated blend of chemicals and other materials whose design and composition exceed the technologies we have currently available.

## The killer app for 3D printing

3D printing food will change the way we eat and how we manage our health. When digital cuisine is as widely accepted as personal computing is today, our refrigerators will hold cartridges of frozen pastes of dark chocolate and pesto chicken. Amateur bakers will download a cake recipe and print out a one-of-a-kind scrumptious pastry whose complexity rivals one made by a virtuoso chef. Home food printers will have settings to allow cooks to select a food's texture, crispness, and perhaps write a custom message inside that will be revealed at first bite. Lovers or family members in distant locations will share a recipe for the same cake which they will print and eat together while they spend time on a webcam.

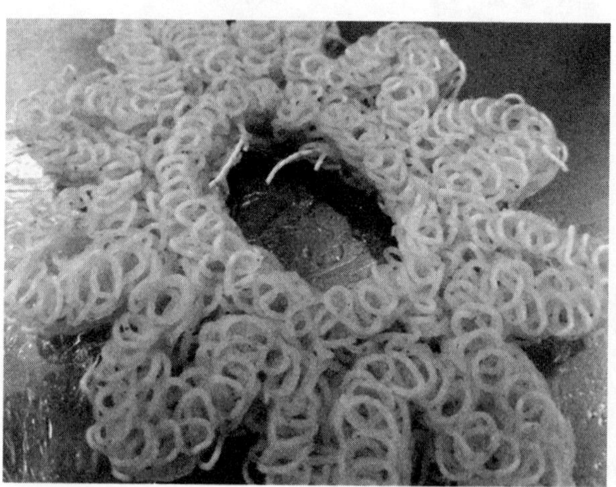

Image courtesy of Jeffrey Lipton and Chef Dave Arnold

**Meandering corn masa printed into a flower shape. Delicious when fried.**

I'm not entirely convinced that every home will someday have a 3D printer. However, if I had to bet that some type of 3D printing technology were to become a mainstream home appliance, I would put my money on food printing. Food printing reminds me of the early days of personal computing, when video games seemed frivolous at first, motivating millions of people to buy their own home computer. Food printing may be to 3D printing what gaming was for personal computers: The frivolous activity that turned out to be the "killer app."

# 9 A factory in the classroom

A few years ago, I was invited to talk about 3D printing at my son's second-grade class. Used to teaching dedicated college-level engineering students, I promptly agreed. "Sounds great," I said to my then-graduate student Evan Malone (now the owner of a 3D printing and manufacturing "gym" in Philadelphia). "How hard could this be?"

In the weeks leading up to the big day, I tested out potential 3D printing presentation ideas on my son at home. Idea after idea was firmly nixed. Would the kids like a presentation on 3D printing? No. How about some movies of printed toys? Or a class on using design software? All negative.

As they day drew nearer and no good ideas emerged, Evan and I felt an unexpected stirrings of stage fright. Finally, inspiration struck: how about we 3D print in playdough? Kids understand playdough, right?

With a viable game plan in mind, we designed a toy space shuttle whose body would be printed in red playdough, and the wings and tail in blue. The day of the demo finally came. We arrived at the classroom with Fab@home, a small open source 3D printer model we had developed in our research laboratory. Fab@home is about the size of a microwave oven, with clear plastic walls so people can watch the printing process in action. For this particular demo, Evan had set up the 3D printer with an extra-large print nozzle so the toy space shuttle would come out faster, in about three minutes or so—within the attention span of the young kids.

Like a minor royal holding state, Fab@home was placed onto a tall platform in the second-grade classroom. Its print cartridges loaded with red and blue playdough. The printer sat grandly, surrounded by a cluster of second graders peering eagerly into its Plexiglas case. After a few introductory words to the class, Evan hit the "print" button. Fab@home's 3D print head began to zip back and forth, slowly squeezing out red and blue playdough in the shape of a tiny space shuttle as directed by the computer design file.

The kids were riveted.

Some second-graders bobbed their heads in rhythm to the moving print head. Others hummed along as the printer's motor changed pitch. Finally, after a few minutes, a mini playdough space shuttle emerged and was plucked out of its print-bed and held aloft so students could take a closer look.

The reverent silence dissipated as students jockeyed for position with their hands raised. One second-grader asked whether we could change the wing shape and print the space shuttle again. Another student asked whether we had brought more playdough in different colors, to print more shuttles. A third student, grasping a can of playdough pulled from classroom shelves, offered a rough estimation of how many space shuttles he could print from a single can. Another student of an entrepreneurial bent calculated how much profit she could make if she sold each space shuttle for $5 and bought raw playdough at current market rates.

In the words of Dave White, a design and technology teacher at the Clevedon School in the United Kingdom, "If you can capture students' imagination, you can capture their attention."

## Make to learn: Children's engineering

Imagine that you're a fourth grade teacher. You are wondering whether design software and 3D printing can help you teach students basic physics and math. You know you need to impart some core concepts, such as the notion of kinetic energy and some simple mathematical ratios. True, you can teach all of this from a number of tried and true basic math and science lesson plans. But you'd like to try something different, just to see what happens.

Most of us don't realize that what's taught in public schools is the work of a thousand pairs of invisible hands. An elementary school teacher does not just dream up lesson plans that she thinks her students will like. Instead, at least in the United States and many Western countries, classroom curriculum is a living, breathing microcosm of the larger world outside the walls of the schools.

States create education standards. A blend of experts and commercial publishers create some of the curriculum. And parents, administrators and school boards have the final say in what's good education. If you're the teacher, where do you begin?

One possibility could be to adopt the approach taken by a pilot program called Fab@school. Fab@school helps teachers create curriculum that integrates design and 3D printing to teach core math and science concepts. The program's goal is to create child engineers, to get youngsters excited about science, math, and especially engineering and design. The pilot is funded by the National Science Foundation, Motorola, and the MacArthur Foundation. Glen Bull, a professor at the University of Virginia's Children's Engineering Group, is leading the project.

The project's lesson plans are developed and tested by a working group of professors, teachers, librarians and graduate students. So far, about 350 fourth and fifth-grade students and 10 teachers have road-tested the curriculum.

Image courtesy of Glen Bull, University of Virginia

**Two elementary school students work with a Fab@school 3D printer. The printer was modified to allow printing in playdough, as well as cutting foam and other digital manufacturing operations.**

Glen and his team created a lesson plan called "Make to Learn." The lesson features a story about an African boy named William. William's village doesn't have electrical power so he figures out how to make his own wind turbine using leftover tractor parts, scrap metal, and a book about energy. Students will follow William's story and design, test, and 3D print their own plastic wind turbine that when placed in front of a classroom fan, lights up an electronic circuit.

In 20 or so brightly colored pages featuring pictures and clear instructional diagrams, students work through a series of hands-on experiments that introduce them to abstract concepts such as kinetic energy, electrical currents, and gear ratios. In the Make to Learn lesson plan students make the windmill blades on a computer-guided paper cutter. They will fabricate the wind turbine's plastic gears and base on a Fab@home printer.

Students work in pairs. The lesson begins:

> *You have a lot of things plugged in at your house—a refrigerator, a TV, lamps, and maybe even something cool like a computer or an iPod. The electricity to power those devices comes from power plants, which can be powered by wind, water, burning coal, and even nuclear fission.*

The story continues:

> *But, did you know that out of the 7 billion people in the world, almost one quarter of them don't have electricity? That's about 1.5 billion people! If any of these people want light at nighttime, for example, there's not much they can do about it. Unless, like a boy named William, they try to make their own electricity.*

Like any inventor would tell you, in the story, William tries, fails, and tries again. Finally, William finds the optimal design and in the end of the story, his "turbine started to spin—first slowly, then faster, and faster—electricity began to flow. Soon neighbors could see William's house, lit by four electric lights, even in the darkest night."

In the classroom I visited, I watched two girls test their 3D printed wind turbine in front of the fan. The girls informed their teacher that their wind turbine wasn't working; they discovered a problem with their circuit board. The girls explained that they took their windmill apart and figured out a wire was connected to the wrong place. This time they were trying something else. Success! Their tiny light bulb glowed as their turbine's blades picked up speed.

Image courtesy of Glen Bull, University of Virginia

**This is a completed wind turbine. The gears are 3D printed.**

These girls are applying kinetic energy to make electricity and taking an active role in solving a problem. By exposing students to the design problem at hand, the Make to Learn lesson introduces students to the idea of "design criteria." In the lesson, design criteria are introduced as a "set of rules that you follow so you make things that people can use and are happy with."

I watched some other students master another core engineering concept: revolutions per minute (RPM). One boy told his partner that they needed "to figure out their windmill's RPM." Earlier, their teacher had attached a shiny sticker to each windmill blade and had helped the boys count blade rotations using a small tachometer. Armed with the rotation count from the tachometer, the boys shuffled off to a table to do some simple math to calculate their blades' rotations per minute.

After my visit to a few Fab@school classrooms, I asked Glen Bull why first-hand experience is such a powerful pedagogical tool. Glen explained that just watching somebody else solve design problems doesn't help students master science, engineering, or math. "A paint-by-numbers kit never made anyone into an artist. It's the same thing if a kid just sits there a watches a 3D printer print something out. What do you learn? Nothing."

The beauty of Fab@school's approach to curriculum is that design software and 3D printing are not the focal point of the lesson plan. Instead, they are enabling technologies that help teachers and students gain mastery by applying abstract concepts to solve interesting problems. Now that design software and 3D printers are dropping in price, teachers and students can experience the design and engineering process first-hand. In the future, Glen and his colleagues plan to create several more lesson plans similar to "Make to Learn" and expand the Fab@school program into middle schools.

## High school

3D printers help high school design and engineering students fail more quickly. Wait, that doesn't sound good. However, in product design, engineering, and other problem-solving professions, the faster you fail, the more quickly you arrive at a solution.

Imagine how disastrous it would be if a group of civil engineers, for example, did not fail early and instead discovered that one week after the grand ribbon cutting ceremony, their design for a suspension bridge contained a fatal flaw. Lives would be at risk as the bridge cracked and collapsed. The project would have to be started again from scratch, a costly and morale-damaging prospect.

3D printers help students fail early and safely thanks to their value as rapid prototyping tools. An iterative design process is based on designers testing their design ideas as they evolve, similar to a writer creating successively improved drafts of a book. Jesse Roitenberg, director of the educational program for Minnesota-based 3D printing company Stratasys, explained that students find the company's 3D printers useful since they can catch design errors early on rather than investing time and materials in a one-shot gamble.

When I spoke with him on the phone to learn more about Stratasys's line of classroom printers, Jesse said, "Back when I was taking high school engineering classes, a typical project they would assign us would be to build a bridge out of toothpicks and glue." In other words, prototyping was not part of the design process. Jesse continued, "The problem with this was that you would find out at the very end of your project whether your bridge was going to fail or not. If it did, you didn't get to re-do it to figure out what the problem was, and how to fix it."

Dave White, head of the Design and Technology curriculum at the Clevedon School in the United Kingdom, has taught design and technology to middle and high school students for over 25 years. He jokes that having a 3D printer in his classroom enables students to "chuck out non-functional ideas." Two years ago, he installed a consumer level 3D printer in his classroom.

I learned from Dave that students rarely 3D print their entire project. Instead, usually the classroom 3D printer is a support tool for a larger, more complex product, perhaps parts for a robot or a body of a model race car. Students use the classroom 3D printer to make custom parts they can't find anywhere else.

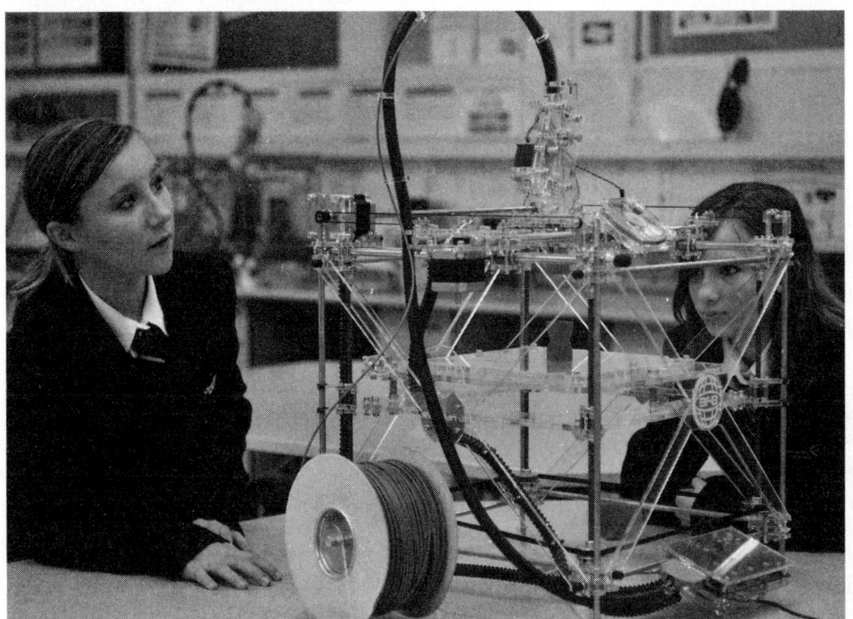

Image courtesy of Dave White, United Kingdom

**Two high school students in the United Kingdom work with a RapMan 3D printer**

Dave's popular design class requires students to design and then 3D print some kind of working object that people can use. Dave explained, "I found that having students make their digital design into the real thing taught them a lot. An object that looks great on the computer screen frequently comes out of the printer looking wrong, or much larger or clunkier than intended."

Dave described the experience of one student who designed an iPod holder that would be clamped onto bicycle handlebars. "On the computer screen, his iPod holder looked great. But then he printed it out and discovered it wasn't very suited for the handlebars of a bicycle—it was too big and not very aerodynamic." The student returned to the software drawing board and adjusted his iPod holder to make it more streamlined and user-friendly.

Dave said, "We have students come up with wild and wacky design ideas that they spend a lot of time making by hand, only to discover too late that their design doesn't work in the physical world. Instead, they should be doing iterative design, testing and improving along the way. Students should design it, print it, try it, and then try it again," he concluded.

Dave works with teachers from other parts of the school curriculum. He explained that other teachers are experimenting with applying 3D modeling software and 3D printers to teach students to design and print DNA structures, models of cells, or replicas of objects of historical significance. A recent project with students in a geology class is a good example.

For a geology class, Dave helped students design a topographic depiction of Mount St. Helens before it erupted in 1980. First, students printed a model of the intact volcano, pre-eruption. Then they printed their topographic models of the volcano after it erupted, complete with a gaping crater in its center. "In arts and humanities courses, students can design and print artistic sculptural forms or historical buildings," he said.

Image courtesy of Dave White, United Kingdom

**A 3D printed model of Mount St. Helens**

Despite the high-level rhetoric about the critical role of design and engineering education, neither the United States nor the United Kingdom have engineering or design skills in their standard curricula. While there is plenty of curriculum dedicated to math and science, there is relatively little attention given to engineering, which focuses on synthesis and design. Therefore, public schools have a hard time justifying funding for the necessary technologies in their budgets. Inside the classroom, teachers must defend their choice to use 3D printing and 3D modeling software in their lesson plans.

In the United States, the Department of Defense and DARPA are funding programs to provide specialized high school and middle school engineering training sites with 3D printers. However, valuable as they are, these pilot programs leave most public schools out in the cold. Public school teachers who aren't funded by a government agency or company must scrape together the resources to procure a 3D printer for their classrooms and get permission to adapt their teaching.

In the United Kingdom the situation is similar. Dave White's 3D printing curriculum is not funded by an external agency, nor does his school have the technology budget to cover it. "Most of the resources I've gathered for my classroom have been gained by begging and borrowing—but of course not from stealing!" he joked. A small company near the Clevedon School called Bits From Bytes (recently acquired by 3D Systems) generously donated a RapMan, a consumer-level 3D printer, to the school.

Dave plans to forge onwards, however. "I spend my own time outside of the classroom working on ways to teach with these new technologies because it's what I believe in," he said. "Schools can't teach kids how to be carpenters and plumbers any more. We have to teach them skills that they can use in their future workplaces."

## Not a national crisis . . . but learning should be enjoyable

Every few decades in the United States, in response to a perceived threat to national security, education experts decree a crisis of public education. Science, Technology, Engineering, and Mathematics (STEM) education has once again become a battle cry. In the 1950s, there was public concern that compared to the United States, Russian public high schools graduated twice as many high school students who would go on to study engineering in college and 30 times the number of skilled technicians.[1]

Today, instead of the Cold War era Soviet "Red Menace," the looming economic threat is low-wage manufacturing countries. As the contemporary storyline goes, the U.S. public school system is not producing enough students who will grow up to be scientists and engineers. Massive reports compiled by federal agencies, educators, and economists claim that unless public schools improve the way they teach science and math, our nation will be crushed on the economic battlefield.

The reason I mention this is because frequently the education value of 3D printing is positioned to policymakers, educators, and the tax-paying public as a way to improve STEM education. The pitch goes like this: If public school science, engineering and math curricula were more appealing, more students would elect to study engineering in college and maybe graduate school. 3D printers will help attract students to STEM courses. 3D printing in public schools will result in more trained engineers and technologically savvy people entering the workforce. The result would be a surge in entrepreneurship and technological innovation. More innovation would create new jobs and high-tech companies.

Maybe it's true that 3D printers will inspire a new generation of students to embrace science, math, and technology. I hope so. The problem with a STEM-oriented pitch, however, is that by justifying investment in 3D printing and design as a teaching tool for STEM education, these exciting teaching technologies are at risk of being cordoned off into a confined pedagogical niche.

Instead, design and fabrication tools such as 3D printers can ignite learning across all disciplines, including the chronically under-funded humanities and arts. A few decades ago, computers were considered only relevant to STEM topics. Today, computers have transformed art, writing, history, and other fields outside the sciences and math.

Another benefit of thinking broadly about design and 3D printing is their value as a teaching tool to reach students who don't do well in mainstream education. Not everyone does well in a "theory first" approach to education. Some students struggle to master abstract concepts if they haven't ever seen or touched the subject they're studying.

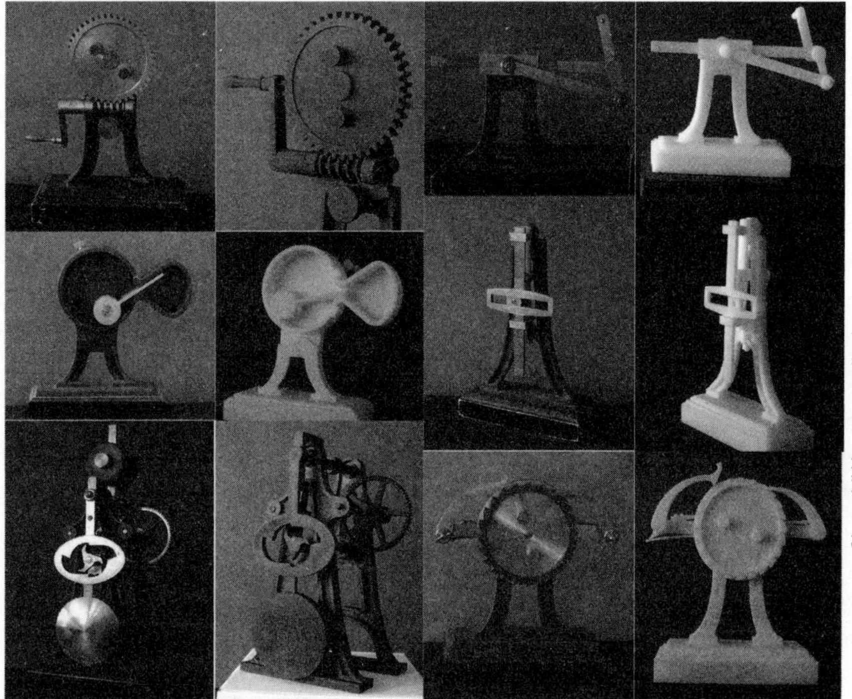

Images courtesy of kmoddl.library.cornell.edu

**A museum of 19th century machines and their downloadable 3D printable replicas**

Visually impaired students may benefit from being able to hold physical models of abstract concepts in their hands. In addition, for all students, regardless of how they absorb new knowledge, re-creating lessons in a physical medium adds another layer of reinforcement to what they're learning. Maybe most importantly, many students enjoy hands-on learning.

Many people underestimate the fact that for too many students, the school day is a long and dreaded exercise. In the words of Glen Bull, "learning should be enjoyable." Shallow as it may sound, if you were a 7th grader, what sort of lesson plans would make you look forward to coming to class every day?

Let's take a look at 3D printing in a purely pedagogical context. Could it be possible that giving students access to tools of design and production could simply improve how they learn? Perhaps (as teachers have noticed) students who don't excel in traditional methods of teaching may flourish when given access to their own tools of design and production.

## Test scores vs. hands-on competence

When I spoke with Dave White, he offered an observation that I've heard from several engineering and design teachers. Over the years, Dave has learned that students who are extremely proficient in their schoolwork, or are what's colloquially known as "book smart," aren't always his star students. "I'm never surprised when I see students who are less passionate about textbook and written assignments come to life once they reach the design portion of a project," he said.

Glen Bull has also witnessed the gulf between mastering a theory and mastering the actual practice. "I've seen kids who just read about electrical circuits create them perfectly in a simulation environment," he said. "However, these same kids, when faced with a real electrical circuit, frequently fail. There's something about the physical dimension, about experiencing learning in the physical dimension along with the theoretical dimension, that changes our brains and makes learning really stick."

Previous research supports Glen's first-hand experience. In 1995, a researcher compared student prowess in making a real electric circuit against their written test scores about the subject. Students in the study demonstrated their practical knowledge of circuits by doing hands-on tasks such as making and testing a circuit. The same students also took a question-and-answer test about how electrical circuits work. The results of the study indicated that students who tested well did not demonstrate mastery when faced with real circuits. In other words, written test scores were not a reliable predictor of a student's ability to work with a circuit in the real world.[2]

Research raises the question of what is more effective: to teach students new information by first exposing them to the theory and then later to application? Or does learning bloom more brightly if students first experience the knowledge first-hand—by doing it—and then diving into the theory? This question is one of critical importance to people trying to figure out how extract pedagogical value by knitting design software and 3D printers into the curriculum.

The answer, obviously, is that each student is different. Different students prefer different things. Even individual learning preferences shift depending on the subject at hand and the complexity of the new knowledge.

## Forget learning styles

We hope that people trying to create curriculum for 3D printing and design software won't fall into the trap of slotting these technologies into specific learning styles. The notion of learning styles holds tremendous popular appeal for teachers, parents, companies that sell educational curriculum, and students. The problem is that the existence of learning styles isn't supported by scientific research.[3] In fact, some research indicates that students pay closer attention, work harder, and do better in classes where teaching is outside their preferred style of learning.

The notion that each of us has an optimal learning style has somehow wormed its way into mainstream wisdom about the classroom and is now firmly lodged in teaching strategy. Somewhat similar to belief in astrological horoscopes, for a few decades now, teachers, parents, even students have embraced the idea that everybody has their own unique style of learning. In response, over the past few decades, scholars of educational and learning theory have identified and mapped more than 70 documented learning styles.[4]

The danger of tailoring curriculum to learning styles, however, is not so much the lack of empirical evidence. The danger is that framing curriculum this way is limiting and distracting. Trying to map new and challenging design curriculum to chimerical learning styles dilutes teacher time and classroom resources.

We should probably explain how researchers define learning styles: The term "learning styles" refers to the concept that individuals differ in regard to what mode of instruction or study is most effective for them. People who favor learning-style assessment contend that the best way for a person to learn is to first diagnose their individual learning style and tailor the lesson plans accordingly, a process known as meshing.[5]

Typical learning style assessments ask people to evaluate what sort of information presentation they prefer (for example, words versus pictures versus speech). Another common question is what kind of mental activity a person finds most engaging or congenial (for example, analysis versus listening). It's tempting to assume that design and 3D printing will appeal to students who have been deemed to have a "visual" or "tactile" learning style. The problem is that this would leave out competent students who, if tested, would fall into a different category.

If learning styles are dubious science, is there any theoretical basis to support the value of 3D printing and design tools in the K-12 classroom? Maybe. People who have a firm mastery of a piece of knowledge or an abstract concept can re-create a concept in a new medium. A more scholarly way to describe the same knowledge depicted across different mediums is that it's presented in "multiple representations."[6]

Here's an example: When you think of the term "parabola," what comes to your mind? If you truly understand the concept of a parabola, you will at once think of a polynomial equation, or perhaps a geometric curve, a concentrating mirror, and maybe even the surface of a spinning liquid. These multiple representations—analytical, geometrical, optical and physical—all relate to the word parabola and a person with true mastery will see the connection between them all. But if you thought of only one of these, then you probably haven't yet mastered the concept fully.

One of the most powerful aspects of transforming a digital design into a real, physical object is the act of reinforcing abstract knowledge into different mediums. Knowledge contained in a design file is virtual, captured in a digital medium. Knowledge becomes physical after it's printed. Design tools and 3D printing shine as tools that challenge students to depict a concept or process in a new medium.

## Now let's see you draw that abstract equation on a graph

People are familiar with the value of using a visual medium to clarify a complex abstract topic. Many of us are equally comfortable with the notion that the best way to enforce new knowledge learned from a book is to apply it to solve a practical, real-world problem. One reason 3D printing is a valuable educational tool is that it brings an additional physical dimension to the teaching and learning of abstract concepts.

For example, in the Fab@School's "Make to Learn" lesson, kinetic energy was defined as "the energy possessed by a system or object that is the result of its motion, and that increases as the object gets faster, or is larger." Such a written description is one way to present the concept. An additional medium to reinforce the concept was provided when students designed and made a windmill and placed it in front of a fan to light up an electronic circuit.

When Dave While led students in a geology class through the 3D printing process, they journeyed through several different mediums. First, students studied data that numerically depicted the shape of the mountain's surface (data is one medium to describe a volcano's shape). One dataset depicted the volcano's shape before the eruption, another after. Next, students translated the raw topographic data into a working design file (digital is another medium). Finally, students printed out mini-replicas of Mount St. Helens after it erupted (the physical medium).

We'll probably never know, but I wonder whether 10 years from now students will remember this geology exercise. What will be remembered? If some students had only read about the exploding volcano while others proceeded through the entire exercise, I wonder which group would better remember the lesson plan.

Here's an interesting experiment to test the power of multiple representations. First, write down a simple equation, say $y=1/x$. Written mathematical equations are one way to depict a relationship between numbers. A second medium is visual, drawing the equation on a two-dimensional graph.

Now let's image that you were to take this simple written equation to the mall or your state legislature and ask several adults to solve the equation for the value of "$x$." Probably several people would know how an equation works. They would know that to solve for $x$, you must be given the value of $y$ (or vice versa). If you were to tell them that in this case, the value of $y$ was 2, they could fairly easily calculate that the value of $x$ would be ½.

How about if the same written equation were expressed in a new medium? Let's assume you asked a person who successfully solved the written equation to express its shape by drawing it on a graph. If you returned to the mall or state legislature and asked random people to do this, probably very few people would be able to readily express the written equation in a chart. Most of us don't retain our high school or college math skills, one reason that among applicants to MBA programs, the acceptable range of math GRE scores for senior-level executives is much lower than the acceptable range for applicants fresh out of college.

Some educational theorists speculate that the reason most people forget their math lessons quickly after they leave school is because they learn math in a single abstract medium. If students learned abstracts mathematical concepts and then translated the concepts into different mediums, perhaps they would gain greater mastery and recall. Translating an abstract concept between different mediums requires skill and mastery. In my own experience, rote memorization of abstract concepts got me through the exam but I quickly forgot most of what I memorized after the test.

It's easier to teach and learn concepts in a single medium. Memorization will get you through. However, real mastery hasn't necessarily taken place. The problem is that most classroom education focuses on the page. However, that's not necessarily what most people remember.

Students who participated in the Fab@school project may later remember the written definition of kinetic energy. Those fortunate few with perfect recall will, in fact, be able to recite it exactly several years later. However, most students are more likely to remember kinetic energy as what made it possible for the spinning blades of their wind turbine to light up the electronic circuit.

## Tactile learning and physical manipulables

Mathematical equations can describe a simple two-dimensional line. More complex equations describe three-dimensional objects. If it's a challenge to transfer the same model from written equation to a flat graph, imagine the challenge of teaching a student the interplay of pressure, volume, and temperature (a thermodynamic equation).

Teachers call a three-dimensional teaching aid a "manipulable." Physical teaching manipulables support what's known as "tactile learning." Tactile learning is not a learning style. Tactile learning is a vital learning channel.

In tactile learning, rather than seeing a picture of a diagram on a whiteboard or a computer screen, students hold a three-dimensional model of a core concept in their hands, and using their sense of touch, are able to absorb and process the knowledge. For visually impaired students, tactile learning is a critical intake channel. If students capture a concept in a design file and print a physical manipulable, the shift in medium might help reinforce the students' new abstract knowledge.

A few years ago, I received an email from Professor Creighton Depew at the University of Washington's physics department. Creighton was teaching thermodynamics. Thermodynamics is the study of the interplay of the effects of pressure, volume and temperature in closed systems. For example, if you've operated a bicycle pump, you noticed that as you compress the handle the pump gets warm. When you compress air, you increase its pressure and decrease its volume, therefore its temperature rises. There's an ongoing, dynamic three-way interplay between pressure, volume, and temperature.

While most students could see the P-V-T (pressure-volume-temperature) diagrams Prof. Depew showed in his class, one student, Dave Plassman, who was visually impaired, had trouble visualizing their complex spatial relationship. Thirty years ago, to help Plassman grasp the basic principles of thermodynamics despite his blindness, Dr. Depew hand-shaped a clay model to represent relationships that other students were able to see on the chalkboard. The clay model was roughly 4 inches high and 6 inches across its base.

Dave could slide his finger along the direction of increasing pressure and "feel" how at each point, the surface sloped downwards in the direction of volume, suggesting that when pressure goes up, volume goes down.

Image courtesy Glen Bull, University of Virginia

**Tactile manipulables: 3D teaching manipulables made from folded cardboard, and then 3D printed**

Thirty years after graduation, Dave remembered the clay model and emailed Depew to ask whether it was still in existence. As luck would have it, yes, the clay model had indeed survived the test of time. Since Dr. Depew was retiring from the University of Washington, after he cleaned out his desk, he shipped the original clay model of thermodynamic principles to my lab. Both Dave and I wanted to see whether we could scan the original handmade model and turn the scanned data into a design file that other physics teachers could use.

The clay model arrived in the mail but was shattered into pieces. After careful reassembly, we successfully scanned and converted it into a design file, which is available for free on the Internet. Today anyone with access to a 3D printer can re-create an exact replica of Dr. Depew's formerly one-of-a-kind handmade clay model of thermodynamic principles. I don't know if anyone has ever used this model and if any other visually impaired student benefited from it. But the opportunity is there.

**Original (left) and 3D printed replica of the clay model of thermodynamics PVT**

At the physics department where my father used to work, I remember seeing hallways laden with old, dusty teaching models of crystals, pendulums, springs and ratchets. These models were beautiful, clearly made by someone with love and care for teaching. Yet that care and talent was not enough to get the models used by others: once the model's creator retired, so did the model itself. But 3D printing might breathe a new life into these old models. 3D printers enable easier sharing and dissemination of physical models, and increase the incentive for creating them in the first place.

3D printing opens a potential motherlode of possibilities for creating teaching manipulables. Custom-designed 3D printed teaching manipulables give teachers the ability to make unique teaching tools that aren't in a standard curriculum kit. Teachers can share and enhance one another's 3D printed manipulables and adapt them for their own lesson plans.

At the primary school level, manipulables could explain simple concepts to elementary school children. Students could create a three-dimensional model of a rare insect or copies of delicate archeological relics. In more advanced classrooms, manipulables can help students understand complex concepts, like models of molecules or the parts for mechanical instruments.

## Barriers to classroom adoption

Today, 3D printers have made solid inroads at the high school level. If there's already 3D design software and a CNC mill in a high school, it's not that big of a step to install a low-cost, consumer-grade 3D printer. In a sense, when a 3D printer scoots into a classroom on the heels of 3D design software, it's exactly where it should be: a classroom manufacturing tool that can turn design ideas into physical reality.

The primary barriers at the middle and primary school level are training teachers, good curriculum, and fitting 3D design and printing into the standardized testing process. Teachers need to be enthused and comfortable about integrating design software and 3D printing into their core math and science curricula. School boards and parents need to understand its role in the teaching process. And somehow we need to devise a curriculum that integrates their capacity.

Teachers breathe life into their classrooms and they determine what sort of learning activities their students will take part in during the school year. Many elementary school teachers, however, are typically not comfortable teaching math and science. They view themselves primarily as reading teachers.[7]

Somehow, to prepare teachers to warm up to 3D printing curricula, we need to ensure that the printing curricula don't just walk teachers through the basic mechanical aspects of the design and 3D printing process. Not everyone gets excited by new technology for its own sake. Many teachers and students only get excited when they see that they can apply design and 3D printing technologies in their own creative arena or to solve problems in their daily lives.

If public schools are to invest in 3D printers and curriculum, to justify their investment, lesson plans need to align with national and state education standards in order for teachers to commit to teaching these activities. In the 1980s, in the United States, most states approved policies aimed at improving the quality of K–12 education by implementing statewide curriculum guidelines and frameworks. About half of the states require students to pass some form of exit examination to graduate from high school.

Private schools enjoy more freedom since they aren't required to adhere to state standards. In the public school system, however, teachers and schools live and die by their students' performance on yearly standardized tests. Standardized tests are a double-edged sword. On the one hand, underperforming schools can be identified and perhaps offered additional funding or assistance.

On the other hand, ironically, the very standards programs intended to improve the quality of public K–12 education may also act as a barrier to the adoption of 3D printing and design curriculum. Because of the emphasis on test performance due to the No Child Left Behind Act, it has become risky for schools to introduce new content into an already crowded curriculum.

In addition, at most K–12 schools, engineering and design are not a core educational requirement and therefore not part of a standardized test. In order to embrace design and printing technologies, teachers must figure out how to apply them in support of state standards. The challenge lies in translating vague and conceptual standards into crisp, relevant, and interesting lesson plans.

Some states set statewide curricula and approve textbooks for statewide use. Typically, however, the development and use of curricular materials is the responsibility of a local school district or a school. The good news about state standards is that standards tend to be vague, leaving teachers lots of wiggle room to teach what they feel is best.

There is little prior formal research on the best way to integrate 3D printing and other desktop manufacturing systems into classroom instruction. Testing new curriculum is not a simple procedure. The U.S. Department of Education asks that public school teachers engage in what's known as "Evidence-Driven Curriculum Design."

Image courtesy of Ryan Cain

**MakerBot funds K–12 public school teachers to create innovative classroom lesson plans. In a second-grade classroom in Brooklyn, New York, students learned about erosion by designing houses and placing them on a sand "river bank" where they were washed away.**

Inspired by the way that the FDA evaluates the effectiveness of new drugs, new teaching methods or curricula must be supported by data obtained from controlled randomized classroom experiments. Teachers and researchers can't just use some new teaching method because of a hunch they have. Experiments first need to compare the performance of students subjected to the proposed new teaching method against the performance of students subjected to the current standard method (the baseline). To qualify, the new method must demonstrate a statistically significant improvement in both short term and long term performance.

The challenge with Evidence-Driven Curriculum Design is that it is difficult to perform blind, unbiased, randomized experiments in the classroom. It takes time, and everyone involved—the teachers, the students, and their parents—is often aware of which group students are in. If participants have negative preconceptions about a particular method, their behavior will likely be affected, and they won't willingly embrace a method they already disapprove of. Perhaps some schools are already engaging in experiments to test the impact of design and 3D printing on student learning. It will be a slow process, however.

## The road ahead

Is 3D printing going to change the classroom? Yes. Will there be a sudden and dramatic transformation of classroom curricula? No. Like any new technology, 3D printing will move into classrooms in fits and starts, embraced by some schools and subject areas and ignored by others.

The more difficult path lies in elementary and middle schools. Most primary and middle schools don't have a design and engineering curriculum. To weave design and 3D printing into a mainstream public classroom, teachers must somehow attach these technologies to support students' acquisition of traditional, core knowledge. As a result, introducing 3D printing becomes a stretch, one that tests the ingenuity of the teacher.

We should mention here that some of the boldest and most innovative uses of 3D printing take place in college and university curricula. However, it's easier for university professors to experiment with 3D printing in their classrooms. Lesson planning in public K–12 classrooms is less flexible and public school teachers don't have readily available funding to pay for technologies.

It would be unfortunate if design tools and 3D printing follow the path already laid by computers. Affluent K-12 schools and universities are well equipped with fast bandwidth and tech-savvy teachers. In contrast, under-resourced school districts limp forward with a single (usually locked) out-of-date computer room where students pay scheduled visits. Magnitudes more students pass through the public K–12 school system than through any other educational avenue. We believe that 3D printing has something to teach students of all learning capacities and socioeconomic backgrounds.

Looking back at the way computers affected education, it is easy to draw parallels. Computers were initially used only to enhance classes that were deemed "computer related"—like a class on programming, or maybe math class. But that changed, and today computers are used in each and every class, from history to art. Most importantly, they have opened the door to entirely new ways of teaching and learning these subjects, and the end of possibilities is nowhere in sight.

3D printers are likely to follow the same path. Initially, 3D printers have been adopted in classes where they are deemed relevant, like the tech shop class. Soon they will be adopted in other classes, from math and biology, eventually making their way into art, history, and literature. Like computers, they will open the door to entirely new ways of teaching and learning that we cannot yet imagine today.

# 10 Unleashing a new aesthetic

My neighbor designs housewares. When I go to his house for dinner, he shows off his latest best-selling creations. One evening he showed me a wavy lampshade. Another time he passed around an interlocked set of salt and pepper shakers. Regardless of what he designed, once the evening's show and tell began, I already knew where the evening's conversation would end.

My neighbor liked to tell his dinner guests that the design phase was the fun, but just the beginning. The real challenge, he would say, the thing that separated the amateurs from the pros, was getting a design manufactured. Making the leap from a design prototype to mass-produced product was the equivalent of jumping over a gaping chasm.

My neighbor would explain that a good designer had to make sure a design idea could actually be made on a factory machine. A designer had to also be a salesperson and convince a manufacturer that his design would make money, enough to justify the manufacturer's sizeable investment in setting up a factory production line.

I haven't seen my neighbor in a while. But I can't wait to tell him that his dark days are behind him. 3D printers are the output device that designers and artists have been waiting for. Complex, unique shapes may be a cause for concern for manufacturing engineers. Yet, for artists, fashion designers, jewelry makers and architects, complex shapes and novel geometries represent unexplored new opportunities.

Architects, industrial designers, and artists are quickly and eagerly tapping into a vast new reserve of design possibilities. 3D printing is removing barriers of resources and skill that prevented many talented designers from realizing their ideas. 3D printing and design technologies are making their first commercial inroads into fields that do small batch manufacturing: for example, jewelry making, high-end home décor, and experimental fashion design.

## Computers that act like nature

Many objects found in nature have regular dimensions that when measured, map to mathematical equations. Have you seen a conch shell cut in half? the conch shell's spiral is a physical manifestation of an ancient mathematical concept called a Fibonacci series. The curves of its inner spiral always have the same shape, whether the shell is the size of a ping-pong ball or the size of a gigantic melon.

The Fibonacci series is found everywhere in nature. Tree branches fork according to this sequence. So does the shape of ferns and artichoke flowers, and even the patterns in swirls and twists of ocean kelp.

A Fibonacci series proceeds in a systematic manner. Each number is the sum of the previous two so the sequence goes 1, 1, 2, 3, 5, 8, 13, 21, and so on. A simple rule can generate this unfolding sequence. A computer, given this rule, can calculate long series of Fibonacci numbers with ease.

As computing power increases, researchers are finding that one of the most effective ways to mimic nature's design intelligence is to apply mathematical rules, or algorithms, to generate shape. Computer-generated 3D fractal art has been around for a few decades, but until recently, has remained imprisoned in the virtual world. What's changed is that 3D printing is making it possible to pull elaborate abstract models out of the computer into the physical world. Before 3D printing, the inner chambers of a conch's chambered spiral could not be made by anything other than nature.

Designers have used conventional design software for years. Organic design is a new paradigm that's flourishing now that there's finally an output device that can unleash those concepts into physical reality. New worlds of design possibilities are opening up as 3D printers liberate mathematical models and natural laws from their abstract confines.

Patterns generated by algorithms, or equations, come in as many varieties are there are people. Using a combination of data and algorithms, designers can create a broad variety of two and three-dimensional shapes and patterns. Some algorithms generate branching structures. Others create curved shapes such as a mass of soap bubbles. Some generate random angular spikes such as quartz crystals.

Nature's manufacturing process is iterative. Every living organism, from a simple plant to a human embryo, follows a relatively small set of developmental "rules" that apply iteratively starting with a simple germ cell or seed. Like a repeating mathematical formula applied to data, a seed develops in shapes, forms and patterns, driven by cues from its environment or limited available resources. On a smaller scale, a sheet of ice crystals on a car's windshield grow and spread according to a regular repeating pattern, starting from a seed crystal.

Coffee tables can grow, too. The Fractal-T is a 3D printed coffee table, the creation of designers Gernot Oberfell, Jan Wertel and Matthias Bär. The designers describe their stunning 3D printed creation as one whose appearance reinforces "the growing bond between nature and mathematical formulas." The shape of Fractal-T was inspired by the regular structure and growth patterns of tree branches. The table was manufactured as a single, 3D printed piece—no seams or joints—using stereolithography and translucent plastic resin. Its designers point out that "the Fractal-T would be impossible to produce using other manufacturing methods."

Image courtesy of MGX, a division of Materialise. Designed by Gernot Oberfell, Jan Wertel, and Matthias Bär.

**A coffee table whose shape is based on an algorithm that mimics the structure and growth of tree branches**

The Fractal-T is a masterpiece of geometric form. Intertwined treelike stems weave through the table and divide into smaller and smaller branches until they get very dense towards the top. Its semi-translucent body is reminiscent of hardened tree sap. The table has been exhibited at several elite museums, including the Victoria & Albert Museum (V&A) in London, the Metropolitan Museum of Art in New York, and the Design Hub in Barcelona.

If you browse the Shapeways website, 3D printed jewelry is one of the most popular items on the market. Jewelry is relatively small. The fact it's not mission-critical (meaning lives don't hang in the balance the way they do in a 3D printed aircraft part) makes jewelry a popular thing to design and print.

One of the most well-known and earliest designers in the 3D printing space is Bathsheba Grossman. Bathsheba designs sculpture, jewelry and housewares whose geometrics represent famous laws of mathematics and physics. Bathsheba's designs have a common theme: math.

Image courtesy of Bathsheba Grossman

**This 3D printed piece by designer Bathsheba represents a mathematical concept called a "Klein Bottle" that she adapted to be a bottle opener.**

On her website, Bathsheba writes "I'm an artist exploring the region between art and mathematics." A typical design is the physical capture of a famous mathematical algorithm, the Borromean Rings, in a pendant made of three interlocking rings whose edges never touch. Like the Borromean Ring pendant, most of Bathsheba's creations have intricate and repeating inner geometries that must be fabricated on a 3D printer. Traditional jewelry making methods such as wax molds or welding can't create repeating hollow inner chambers.

Closer to the sub-$100 price range of most consumers, designer Unellenu specializes in designing 3D printable fractal sculptures and jewelry. Unellenu's Shapeways storefront has several pages of offerings such as branching pendants in different metals, each representing a mathematical model. The Pythagoras Tree is a jewelry holder; its algorithm-driven design of spikey, triangular branches is a hanger for necklaces, earrings and other finery.

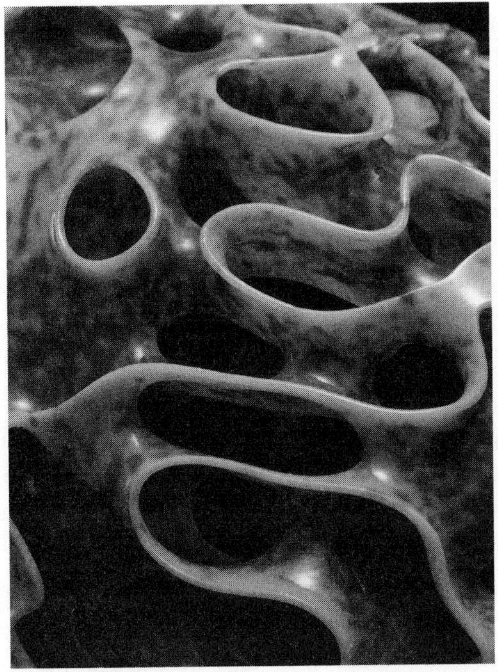



**MIT professor Neri Oxman is one of the leading researchers in generative design. She uses 3D printing to fabricate shapes designed by mathematical algorithms to make forms that would not be possible to make using conventional materials such as glass, steel, and wood.**

Nervous System, a generative design company founded in 2007 by MIT graduates Jessica Rosenkrantz and Jesse Louis-Rosenberg, designs the mathematical models that drive the designs of their 3D printed housewares and jewelry. Their wares, stylized shiny renditions of algae, coral, cells, and veins, are inspired by laws of nature. Nervous System shares their mathematical systems in an interactive software applet with customers, who can apply the system to design their own unique products.

Josh Harker, an independent artist and designer, describes himself as a classically trained artist and sculptor who "uses bits, ones and zeros, to express himself in a human way, to make something new." I stumbled upon Josh's artwork on crowdfunding site Kickstarter where he raised a record amount of funds for a sculpture project called Crania Anatomica Filigre, a white, plastic ornately filigreed 3D printed skull. When we spoke on the phone, I learned that he grew up in the Mississippi River valley in Illinois. He described his bohemian childhood as one that "included post '60s off-grid communal living, Hell's Angels babysitters, complete artistic immersion, and family tragedy." Today he's a full-time artist and a leading digital sculptor. Josh explained, "Most of what I do is digital."

Josh learned how to use design software and 3D printers several years ago when he owned and managed a boutique design firm. "I don't know how a designer could stay in the marketplace these days without knowing how to use the tools," he said. After nearly a decade in industry, the call to return to art and sculpture became too strong to resist. In 2008, Josh returned to art full-time to explore his passion for digital sculpture and 3D printing.

Josh initially embraced 3D printing to fabricate the elaborate geometries he creates on the computer. "I used to create geometries on the computer that were too complex to make," he said. "I did drawings for years but they were too complex to hand sculpt. Clay, wood, stone—nothing would work.".

Some of Josh's sculptures reflect his classical training—nudes and character studies. Many of his sculptures are digitally designed and 3D printed. His "Knots & Tangles" series reflects his fascination with the design potential inherent in natural formations such as roots, vines, neural networks, or cardiovascular systems.

"I'm lucky to be around at the time all of this is happening," said Josh. "It's a revolutionary time for the arts," he added empathically. "Never before have forms of this organic complexity been able to be developed and reproduced. I look forward to seeing what's ahead."

Inorganic, physical processes can generate shape as well. Eyal Gever of Tel Aviv creates what he calls "Disaster Art." Using simulation algorithms, he reconstructs transient tragic moments that would normally disappear in a blink of an eye. Using a 3D printer, he can bring these moments back into reality, in all their force: for example, a car crash or an oil spill. The effect is that of freezing time in physical reality.

**Oil Waterfall. Fluid simulation of oil was used to compute geometry, then "frozen" to create printed model (inset).**

## Fashion: Optimized soles and sustainable footwear

One of the most powerful applications of generative design is to apply computer algorithms to a particular problem to find the best, optimal solution. By crunching through rapid iterations and testing out possibility after possibility, a computer can generate design specs that when 3D printed, will create an object optimized to suit a particular person or environment. Fashion design is a field ripe for 3D printing optimized, organic designs. We visited the London College of Fashion to learn more about 3D printed couture shoes.

The afternoon meeting at the College of Fashion began when Philip Delamore and several students walked in, introduced themselves, and placed two shoes onto the conference table. Philip explained that both of these printed shoes were the result of a master's student design project. Both shoes were 3D printed, the fruit of a collaboration between the College of Fashion and London-based software design company called Within Technologies.

Philip directs the College's Fashion Digital Studio where fashion students experiment with advanced technologies, such as a full-body scan chamber and more recently, 3D printing. "The fashion design industry is becoming much more involved and interested in the function of fashion," said Philip.

Comfort aside, the patterns looked nothing like any shoe I've ever seen before. I'm not a fashionista by any stretch of the imagination, but these shoes were something different altogether, a clear departure from millennia of traditional designs. One shoe was black, bristling with short quills that resembled a porcupine's back or perhaps a bed of nails. The designer, Ross Barber, told us that a few weeks earlier, these shoes strutted down the catwalk on the feet of a male model.

Ross told us that what made his shoes unique wasn't their looks. It was the engineering and design work that went into them. The company whose software helped design the shoe—Within Technologies—creates design solutions based on nature. For example one of Within's recent projects involved optimizing the heat exchange of a metal engine block. The company's engineers studied the flow of water over the gills of fish to create an extremely efficient engine block whose inner geometries were wavy and gill-shaped.

Image courtesy of Ross Barber

**A 3D printed shoe designed by Ross Barber at the London College of Fashion and optimized by Within. The inner leather shoe was hand stitched in to make the shoe more comfortable when it was worn on the catwalk.**

Ross pointed out to us that there was no glue holding his 3D printed shoes together. They were printed in a single piece out of an extra-durable polymer material that's usually applied to making industrial parts. To improve the shoe's comfort factor, Ross hand-stitched in a leather boot upper from a men's shoe shop. The result, a hybrid foot covering with a standard shoe top, laces and tongue nested inside the 3D printed casing.

Philip passed the shoe around the table. I hefted it in my hand, marveling how light it was. Despite the shoe's appeal as an attention-grabbing fashion statement, it was a marvelous application of durable, engineering-grade polymer and Within's sophisticated computer algorithms.

Anthony Ruto, Within's Chief Technology Officer, was also visiting the College that afternoon. Anthony explained that he and Ross adapted one of the design algorithms that Within's designers ordinarily used to make custom orthopedic medical implants such as artificial hip joints. To meet the design challenge to create a shoe that would be light in weight but durable, Anthony and Ross adjusted the orthopedic algorithm to optimize the structure and shape of the shoe soles.

After a few months of work and trial and error, Anthony and Ross designed and printed shoes that were attractive to look at, were light in weight, comfortable and durable. Anthony pointed out that the shoe's soles embodied the generative design process in action: the soles were formed in a lattice-like structure. This custom-design lattice was what enabled the shoes to be light in weight, yet more durable than a solid, factory-issue rubber sole.

Many of us, when we hear the word "lattice," imagine a rose garden or green, flowery vines climbing upwards on a white, grid-like structure. That's a two-dimensional lattice. Two-dimensional lattices are easy to produce using conventional manufacturing methods. However, until 3D printing appeared on the design scene, three-dimensional lattices were all but impossible to produce.

Lattices are a basic engineering structure that's adapted for a broad range of applications, similar to the way simple raw dough serves as the foundation for elaborate pastries. Lattice structures are a classic example of generative design. They're designed by a computer algorithm that automatically generates a regularly repeating, semi-regularly repeating or random structure. Some three-dimensional lattice structures have an angular internal geometry formed from tiny, repeating regular shapes (such as three-dimensional triangles) while others look like random fibrous tangles.

After Ross finished his presentation, we turned to the second shoe on the table. This shoe, also 3D printed, was an open-toed woman's sandal with an astronomically high off-white wedge heel and a smooth and peach colored upper. This shoe was designed by student Hoon Chung, also under Philip's mentorship. Hoon explained that his design project took a different approach, focusing less on the internal structure and durability of the shoe and more on its market appeal and manufacturing process.

Image courtesy of Hoon Chung

**A 3D printed modular shoe designed by Hoon Chung**

Hoon, like many designers, liked the fact that 3D printing his design concepts gave him more control over the manufacturing process. Earlier in the semester, Hoon had first considered injection molding the shoes. He visited a shoe factory in Spain and witnessed the environmental damage inflicted by glues and other chemical processes the factory used.

Hoon turned to 3D printing and found that the additive manufacturing process enabled him to create a shoe out of modular components. The shoes' parts are interchangeable. So their wearer could snap apart her current pair and swap out the heels for some of a different color to create an entirely new look. Modular, snap-apart parts also don't need to be glued together, a plus for environmental reasons.

Image courtesy of Jenna Fizel and Mary Huang, Continuum Fashion

**Next generation beach wear: a 3D printed bikini designed by Jenna Fizel and Mary Huang of Continuum Fashion**

## Hacking biological data

For most of the recorded history of design in the Western world, people have viewed nature and technology as polar forces. Nature was viewed as random and unpredictable, a force to be tamed. Design was a process of making order from nature's chaos, placing the imprint of human rationality onto raw materials. Think of ornate Baroque furniture styles or the stunning and painstakingly detailed gilded icons found in Russian Orthodox churches.

True, modern architecture embraced organic, nature-inspired shapes decades ago. However, not everyone would agree that an architect's creative role as a designer should be to guide and orchestrate the application of biologically based algorithms to a design problem. That's what makes generative design controversial and disruptive.

If "generative design" is the process of forming 3D shapes and patterns using iterations of mathematical rules and constraints, then is the computer or the human making the design? Some may argue that generative design, since its results are the product of iterations of mathematical models, is not actually a design process—the computer is just blindly running formulas and data through their paces. However, it is the artist who designed the rules.

An evolutionary process "designed" the conch shell's shape to survive a tough marine environment, a process that lasted millions of years and involved billions of design iterations (or new generations). With computing power, data and 3D printing technologies, designers assume the role of the evolutionary process. Designers can automate a set of design rules and watch the resulting shapes form and unfold into a pattern. A designer can apply the same rule to repeat a design process many times. A designer can vary design rules, selecting rules that generate better looking, or more interesting results. New technologies elevate designers, artists and architects to a new role, to a new way of design thinking.

Jenny Sabin is a professor of Architecture at Cornell University and a pioneering architectural designer. Her research explores the intersection of biological systems, computation, and the design of material structures. We interviewed Jenny to learn more about the generative design process and architecture. "In recent years, design processes in architecture have seen a shift from compositional design to generative design," she explained. "Writing and hacking software has become part of the design process."

Jenny graduated from the University of Pennsylvania with a master's in architecture, where she taught for 6 years. While studying for her degree, she found herself drawn to the unpredictability and mostly unexplored creative terrain of computer-based, generative design and 3D printing. "One of the things that's exciting about generative design is that you never know until the end exactly what the final result will be," said Jenny.

One of her current collaborative research projects is called eSkin, which uses cellular data to inspire designs for responsive building materials. She and her team utilize 3D printing to explore and capture biological behavior in component-based generative models. We asked Jenny to help us understand the role of generative, biologically inspired design in printing next-generation building materials.

"The larger agenda for the eSkin project is to understand how buildings may become more biological in the way that they interact and adapt to their environments," she said. The scientists within the eSkin collaboration provide vast amounts of data collected from human cells plated on engineered substrates designed by the material scientists on the team. The architects on the team then try to "reverse engineer" the observed behavior—they design "algorithms" to simulate the cellular behavior and process that generated the observed structure. The team then applies those algorithms to generate new possible design solutions at a larger scale.

The eSkin research is based on the belief that cellular activity can help people learn to design more energy efficient building materials that like cells can respond to changing environmental conditions. In an email, Jenny explained,

> *Cells know how to respond and adapt to environmental changes. We will capture their movements, then digitize them and eventually output the result using a 3D printer. The 3D printer allows us to explore biological behavior and process through the constraints of material and fabrication in real time. We use the 3D printer less as a representational device, but instead to explore part to whole relationships in the form of adaptable components.*

Image courtesy of Jenny Sabin

**A 3D printed object whose design is inspired by cellular structure**

The eSkin research project belongs to a family of approaches known as biomimicry. Biomimicry is described by American natural sciences writer Janine Benyus as "the process of learning from and then emulating life's designs. It's innovation inspired by nature." Much of the excitement around computer-generated design and 3D printing is the hope that finally we will be able to 3D print objects whose shape is optimized for their environment or application.

As technology advances, we continue to return to nature for inspiration. Nature's designs represent elegant and time-tested solutions to the challenges of the physical world. As expressed by design architect Michael Pawlyn in a TED talk in 2010, "You could look at nature as being like a catalog of products, and all of those have benefited from a 3.8-billion-year research and development period."

## Responsive design

In another architectural college in central London, students 3D print honeycomb structures and futuristic dome-shaped architectural prototypes. I stopped by the Architectural Association School of Architecture during a visit to London. The school was well-concealed inside a several grand, gray stone rowhouses, just a few blocks from Trafalgar Square. I learned later that the school purchased these rows of houses years ago, before London housing prices became as astronomical as they are today. From the street, the only hint of the school's presence is a discreet banner hanging from one of the gray stone houses.

Earlier over email, I had been introduced to the School's professor of digital prototyping, Jeroen van Ameijde, who was to meet me for an interview. Inside the school's front door, a hallway led back to a bustling and sunny reception area. The day I was there, the place was flooded with eager prospective students dropping off their applications, precious design portfolios wrapped in brown paper.

Jeroen van Ameijde is one of the resident experts on 3D design and 3D printing. After our introductions, he led me from the reception area down the stairs to the Digital Prototyping Unit, a warren of connected, cavernous underground rooms tucked away in the back of the school. Several small manufacturing machines covered the surfaces of tables and abstract, 3D printed sculptures hung from the ceiling. Most of the area's horizontal surfaces were cluttered with oddly shaped 3D printed design experiments.

The unit's chambers weren't always so educationally oriented. Jeroen told me that these rooms were host to a Pink Floyd concert in the 1970s. Before that, in the 1950s, architectural students used to hold swing dance parties where the manufacturing machines now stand.

Jeroen was hired by the school 5 years ago to teach digital prototyping and design. "I was overwhelmed with interest when I got here," said Jeroen. "There was a tremendous pent-up demand for more computer-intensive design curriculum and it seemed to make perfect sense to people here that the 3D printer was simply an extension of computing as a key design tool."

Architecture has historically had two major schools of thought: those that believe architectural design to be an art, and those that believe it to be an analytical process more closely aligned with engineering. "CAD had already been a part of the curriculum for several years," said Jeroen. "What was new was the idea that the computer had a role in the actual design process, that we could apply algorithms to help us design and create novel structures."

"About 2–3 years after I arrived, digital prototyping permeated laterally into every curriculum here," said Jeroen. "People realize that the design software and 3D printers are not an end in and of themselves. Instead, they're just another tool to support the curriculum they already intend to teach."

"Computers enhance the architectural design process. I call it 3D thinking," said Jeroen. "Now we're starting to get the technology we need to capture the real complexity of things—the data—then we can start to work with that to translate that into an algorithm and from there, into the designed object."

"The hallelujah moment in a design project is when the data is coming in," said Jeroen, aptly summing up the new and growing role of data in the design process. Ever-increasing amounts of data have become the raw material of algorithm-based "intelligent" design and 3D printing. Data flows into the architectural process from many sources. Laser scanning has become a standard part of the architectural design process these days. Sensors provide another rich and growing source of information.

Jeroen explained that his longer term vision (whose realization lies beyond the scope of just a semester or two, of course) is to design a responsive 3D printer. He envisions a construction 3D printer that could go onsite and print homes and other structures on the spot. This responsive printer would react in real time to data flowing in from sensors. Inflowing sensor data would be rapidly interpreted by intelligent design software that would guide the printer to fabricate structures optimized for the conditions in a particular environment.

A responsive 3D printer would receive constantly updated design instructions and would adjust its printing process accordingly. Responsive design software of the future could continually adjust the construction of printed structures. Intelligent design software could adjust the shape of the structure's footprint in order to make the structure more stable. In addition to guiding shape, responsive and intelligent design software could also guide the materials, or combination of materials, used by the 3D printer.

What would such a printer look like in action? On his laptop, Jeroen played me a concept video showing a demonstration. On the screen, a large "print head" laid down layer after layer to create a hive-shaped dome with wavy walls and a pointy top. Part of the design concept for this particular structure in the video demonstration were walls containing printed-in ventilation shafts.

"The intelligence will be contained in the design process," said Jeroen. "As design software becomes more sophisticated, it will be possible to 3D print structures whose shape will be determined by the nature of its environment." Another way to think about this method would be that the design software would act as a dynamic and responsive electronic blueprint "speaking" to the 3D printer in a closed feedback loop.

I learned that Jeroen is supervising a student project at the school to build a 3D printer large enough to print structures up to several meters in size. Since such a printer is not commercially available, Jeroen and students are hacking a large CNC machine and fitting it with a 3D print head. They're tinkering with the machine to create a print head that can swing and swoop freely in space.

Responsive 3D printing is probably 5 to 10 years from being reality, Jeroen told me. I suspect that the physical construction of a large printer is the relatively straightforward part of the project. Creating intelligent design software that can respond and make good decisions on the spot will take longer.

Unlike a steel girder that's dropped into place by a construction crane, a 3D printed structure could be made of custom blends of construction materials adapted to the stresses and loads the structure is subjected to. Like a growing bone that thickens when weight is placed on it, a 3D printed responsive bridge could be thickened in response to real and simulated stress.

An article on the Fabbaloo blog, which follows personal manufacturing and 3D printing issues, put it this way: "Imagine girders that have strong areas where they need to be strong and light in other areas—or even sparse sections with no material at all. Every piece could be specifically made to provide the best physical strength for its particular purpose in the building at the least cost of materials."[1]

Ideally, 3D printed structures could be made from locally available resources such as sand or dirt. Intelligent design software could blend materials together, a complicated art in and of itself. For responsive 3D printing to become a reality, we would need better materials for construction and sufficient computational power to model and analyze the data very quickly and to adapt the design in real time.

Responsive printers could address social challenges. They could be shipped out to fabricate shelter for mobile populations, for refugees or to help victims of natural disasters. In non-crisis situations, as building materials improve, responsive printers could create eco-friendly and comfortable high-density housing. Responsive, intelligent, on-the-spot construction would be useful for the military and for space exploration.

Portable responsive printers could fabricate low-cost homes. According to Morgen Peck of *Txchnologist* magazine, it can take anywhere from 6 weeks to 6 months to build a standard, two-story 2,800 square foot house in the United States. While a good chunk of the cost is raw building materials, another major source of cost and complications are the construction process and human construction workers.

3D printing and computers lets us learn from natural systems and apply the power of mathematical models to the design process. Designer Hugh Dubberly summed it up: "Design is no longer concerned only with things. Increasingly, design is concerned with systems—and now systems of systems, or ecologies. In a sense, these systems are alive. They grow and co-evolve."

## Printing wavy walls and custom gargoyles

3D printed structures already exist. Today, researchers are 3D printing cement homes using conventional design software and custom-made 3D printers. Like their small-size cousins, construction-scale 3D printers can form shapes that until now have been impossible to make, such as channels to improve ventilation, printed-in structural features and for ornamentation, fantastic flourishes and curves.

Behrokh Khoshnevis, a professor at the University of Southern California, is printing buildings. Several years ago, he and his students started testing clay as a printing material. Today, his early work has progressed to the point where he can fabricate large buildings out of concrete using a 3D printer he calls the Contour Crafting construction robot.

The construction robot works on the same principles as a consumer-scale "nozzle" 3D printer that extrudes raw material through a nozzle print head. The robot traces out a footprint of the structure by squirting a stream of concrete paste, reinforced with fibers. As the nozzle moves around, the lower layer hardens to support the increasing weight of the growing walls.

Image courtesy of Behrokh Khoshnevis, Center for Rapid Automated Fabrication Technologies (CRAFT) at the University of Southern California

**Contour Crafting extrudes fiber-reinforced cement from a nozzle**

Currently, Contour Crafting can build a structure that's about 7 feet high, 23 feet long, and 15 feet wide. Behrokh estimates that his printer can build at the rate of about one square foot of wall in less than 20 seconds. To lay down thick streams of cement lines in seconds, the Contour Crafting robot has two swiveling trowels that smooth out each layer of concrete as it's printed.

Right now, the Contour Crafting technology can't embed vital infrastructure into its raw concrete such as pipes and electrical wiring. Someday, however, more sophisticated construction tools could be added to join the two trowels that aid the printer. For example, an additional robotic "pipe layer" could lay pipes into the hardening cement.

One of science-fiction author Bruce Sterling's famous quotes was, "Just as termites build castles on Earth, robots could erect skyscrapers on the moon." With space travel in mind, NASA has funded several of Behrokh's robotic construction research projects as a possible method for building structures in outer space. Behrokh envisions someday constructing structures on the moon and Mars, and closer to home, providing emergency and low-income housing on earth.

On the other side of the Atlantic, outside of Pisa, Italy, Enrico Dini, an Italian designer and architect, has devised a computer-guided, 3D printing construction method that uses sand and an inorganic binder to create artificial sandstone. Sand is one of the most commonly available raw materials on the globe. Enrico describes his printed creations as smooth and marble-like, cool and hard to the touch. He named his construction printer the "D-Shape robotic building system."

A computer connected to the D-Shape printer runs CAD software that guides the robotic arm millimeter by millimeter over a bed of sand.

The D-Shape printer works by extruding liquid adhesive through hundreds of small nozzles onto the bed of sand. The mixture contains a catalyst the makes the adhesive more quickly bind to the sand. After four passes, the printed layer solidifies. Fresh sand is sprinkled on top and the robotic arm is recalibrated so it can deposit another layer of adhesive on top of the first. It takes about a day for the printed walls to solidify.

According to Enrico's website, the adhesive is strong and can substitute for iron as a structural reinforcement. The artificial stone printed by the D-Shape printer could be made of any kind of sand and it's stronger and cheaper than Portland cement. A 3D printed sandstone structure is faster to build than one made of traditional building materials.

Image courtesy of Enrico Dini and Andrea Morgante

**Enrico Dini prints structures out of a blend of sand and adhesives. The end result is a hard and smooth stone surface. The shapes and curves would be very difficult to hand carve out of stone or marble.**

At the time of this writing, I'm not aware of anyone who's residing in a 3D printed home. However, maybe somebody will take the plunge and decide to take advantage of the design possibilities inherent in a printed residence. Similar to the geometric freedom provided by traditional 3D printers but on a larger scale, a construction 3D printer could print odd and unusual physical shapes. Imagine living in a house shaped like a gigantic stone igloo or pueblo. You could have lavish ornamentation on your roof, even gargoyles in your own image.

Image courtesy of Sungwoo Lim and Richard Buswell
Photo: Agnese Sanvito

**The "Wonder Bench" is 3D printed out of concrete. The curved shapes and hollow interior channels would be impossible to make using traditional cement pouring techniques.**

You can't yet order a blueprint for a 3D printed concrete home. But on a smaller scale, you could start with a simple concrete outdoor bench. A team of researchers at the Loughborough University (now at University of Nottingham) developed a 3D printed concrete "Wonder Bench" by integrating and expanding on principles from Contour Crafting.

The Wonder Bench's printed structure is a hive of internal air pockets and channels that researchers call "voids." Voids can act as sound carriers, thermal insulation, or simply provide open space in which to install additional building components, perhaps the pipes and electrical wires we mentioned earlier. The Wonder Bench provides a glimpse of the potential of 3D printing larger and more complicated structures. A printed structure could contain embedded circuitry, wiring and plumbing. It could be formed with intelligent air spaces, or voids, so additional housing infrastructure can be easily added later.

## The robotic designer

In their classic book on computer graphics *Evolutionary Art and Computers*, authors Stephen Todd and William Latham explained, "The computer does not blindly follow the rules, but uses them to give suggestions, and leaves the artist to make final aesthetic selections."[2] Next generation intelligent design software tools will have the same effect on design.

Someday the creative process will be aided by intelligent and responsive software, or a "Robotic Designer," that will speed up the discovery of design solutions. Today, computer algorithms already provide design solutions that are more complex than anything a human designer could have come up with. In the future, a Robotic Designer—attached to a 3D printer and a data source—will change the role of the human designer, artist or architect.

Future humans will no longer spend time drafting and testing out possible design solutions. Instead, we will provide a Robotic Designer with a high-level concept of the design problem—its goals, it constraints, and its context. The human designer will define the parameters of the desired solution. Rather than being rendered obsolete by the new tools, future human designers, artists and architects will scale new creative heights.

# 11 Green, clean manufacturing

A man in a bowler hat and sunglasses stands on a vast sandy plain under a baking sun and blue cloudless sky. Next to him awaits a gleaming metal contraption the size of a small urban bus stop. On top of the machine are two dark solar panels. Nearby, a small tent of metallic fabric serves as his portable office.

The man crouches in front of what looks like an open briefcase. His hands patiently untangle a nest of brightly colored wires that spew from its insides. Behind him, magnificent sand dunes accentuate his glowing white clothing.

Image courtesy of Markus Kayser

**Markus Kayser uses solar energy to power this 3D printer that captures the sun's light through a gigantic lens to melt sand into shapes. The location is a desert in Morocco, but this scene could take place anywhere there's strong sun and abundant sand.**

When the wires inside the briefcase are arranged to his satisfaction, the man connects the briefcase to the machine. The machine's metal frame is jointed together with spools covered in aluminum foil, reminiscent of the knee and elbow joints of The Tin Man in *The Wizard of Oz*.

As the solar panels begin to stream in energy, the machine's gears begin to grind. A rectangular thick pane of glass, attached to the metal frame by an adjustable arm is a gigantic magnifying glass. Its lens concentrates the sun's rays into a powerful beam of light.

Using the moveable arm, the man adjusts the magnifying glass and aims its beam onto a pan full of sand. He connects the laptop to the machine and sits back to see what happens. The machine's arm moves the lens back and forth. The sand under the relentless glowing focal point begins to bubble and melt. Like a time lapse video of a molten pool of lava, the sand heats up to 1,500 degrees Celsius.

The gigantic magnifying glass methodically traces out the shape of a circle that slowly grows into a three-dimensional shape. The melted sand cools and solidifies and new fresh sand is piled on top. The machine continues its work, melting new layers of sand that fuse together on top of cooled and solidified sand. Eventually, the man pulls a large bowl-shaped object from the sand bed and declares the job done.

A white truck drives up and men in turbans quickly load the machine into the back of a jeep. The mysterious sand alchemist, his bowler hat firmly atop his head and clothing still immaculately white, leaps into the back. As the truck drives off into the distance, the sandy plain bears no trace of the manufacturing process that just took place a few moments ago.

This desert scene was filmed in Morocco. The man in the video is Markus Kayser; the machine is a solar-powered 3D printer, a design project that Markus named "Solar Sintering." In a demonstration of green manufacturing, the solar-powered printer's raw material is desert sand that's melted into glass by the attached magnifying glass.

As Markus described it, "In this experiment, sunlight and sand are used as raw energy and material to produce glass objects using a 3D printing process that combines natural energy and material with high-tech production technology." I watched this video clip several times, mesmerized by the soft muffled hush of the vast desert, the bubbling sand, and the merciless sunlight.

Solar Sintering is ecologically ingenious. It runs on solar energy; the printer's "laser" is concentrated sunlight. Sand, its raw material, is one of the most naturally occurring and abundant substances on the planet. When sand is melted, it becomes glass, a strong and versatile material that doesn't need additional chemical additives or glues. If a printed glass object were discarded and left behind in the desert, it would eventually come full circle and be ground back into sand.

When I watched the video, I realized that Solar Sintering could be the answer for a problem I used to puzzle over when I was a child. I grew up in Israel, a country of abundant sandy plains. Building roads on top of desert sand is expensive and difficult; sand doesn't provide a stable base and as it shifts and blows around, roads get covered up and eventually become impassible.

When I was a child, I wondered why the government didn't just melt the desert sand into hardened strips that cars could drive over, a sort of "glass road." Imagine if Markus's Solar Sinterer were equipped with gigantic wheels, attached to a GPS, and put to work printing roads in desert areas. Perhaps a human "crew boss" could oversee roaming road crews of Solar Sinterers. Shifting sands would no longer be a problem. Instead, loose sand would serve as a useful raw material, a glass tar. NASA is exploring similar processes for 3D printing structures on the moon from lunar sand.

In a perfect world, all manufacturing would be as ecofriendly as Solar Sintering. Unfortunately, most manufacturing runs on petroleum-based energy. Factories and global transportation networks (called supply chains) leave behind an enormous carbon footprint which contributes to the build-up of greenhouse gases.

The manufacturing process is just part of the problem, however. Garbage is another environmentally disastrous side-effect of mass production. Cheap and abundant mass-produced products are made, literally, to be thrown away. The problem is that there is no "away." Glass and paper can be (and are) recycled. However, most mass-produced plastics are not recycled, nor do they biodegrade peacefully back into the environment. Disposable plastic products linger for decades, filling landfills and polluting our oceans.

3D printing technologies offer us a cleaner and greener way to manufacture things. Yet no technology is innately green. What matters is how it is put to use.

## A tale of two plastic toys

If someone handed you two plastic toys of about the same size and weight—one mass manufactured and one 3D printed—would you be able to guess which toy's manufacturing process caused less environmental harm? Many people would jump to the conclusion that the 3D printed plastic toy was greener than the mass produced action figure. Would you agree?

The answer lies in each toy's product lifecycle, or the winding chain of production, distribution channels and stores that carry a product to its ultimate destination—the point of sale. The tail end of a product's lifecycle is its disposal, when its human user throws it away.

Imagine a hypothetical scenario where both toys could talk. In this scenario, you ask both plastic toys to tell you where they were made, how they were assembled, and finally, how they reached the person who bought them. Let's imagine that the mass-produced toy was first to answer your question.

This toy would inform you that it began life as a loose bunch of plastic pellets the size of fish eggs called nurdles. Sometimes rogue nurdles escape their packaging and wash into oceans and rivers where they worm their way into the nutritious zoo plankton base, choking and slowly poisoning marine animals and sea birds. This toy's nurdles were successfully fed into an injection molding machine and forced into the shape of the toy, which was dumped onto an assembly line.

The assembly line was probably located in an offshored factory in Southern China where an estimated 80 percent of the world's toys are produced.[1] The mass-produced plastic toy came into being in the company of thousands—maybe millions—of identical plastic toys. Its first glimpse of a human face was likely the factory worker who snapped its plastic parts together.

Mundane as mass-produced plastic toys may appear, most are cosmopolitan world travelers. This toy and its legions of identical colleagues left their factory of origin in a shipping box and embarked on a journey of several thousand carbon-emitting miles over oceans, rails, and roads. The end of their journey was a loading dock behind a store at the mall where a local employee unpacked the boxes. After a few weeks sitting in the store's inventory, the toy was eventually placed onto a shelf to await its sale and final destination: a child's eager hands and home.

Let's imagine that the hypothetical scenario continues and the 3D printed toy takes its turn. The 3D printed toy would explain that it is the only one of its kind. Its design was not dreamed up in a toy company's marketing or design department. Instead, its design was based on a video game avatar.

The 3D printed toy is brightly colored. Like the video game avatar that inspired it, the printed toy boasts an elaborately folded cape that ripples over its back and shoulders. The printed toy started life not as a bunch of nurdles, but as a pile of plastic powder. Eventually, it came to "life" when a customer placed an order on the company's website, uploaded her game avatar, and punched in her credit car number.

When the toy company received the order, company engineers adapted and adjusted the uploaded digital file into a printable design file. Much of the formatting process was automated since the avatar already existed in digital form. Finally, the customer approved the final design file and a small print shop nearby manufactured it.

At this point in the story, the 3D printed toy would describe its first human, a print shop employee who pulled it out of its print bed. This technologically-skilled midwife dusted off the toy's excess powder and buffed and polished it to perfection. Finally, another print shop employee placed the completed toy into a small box and FedExed it to the customer's front door.

Which of these two plastic toys has the greener product lifecycle? At first glance, it seems the 3D printed toy is more eco-friendly. It was made in a clean, regulated print shop in a developed nation where working conditions are good, safety standards are met and labor regulations followed. Shipping one small FedEx box leaves a smaller carbon footprint than than shipping hundreds of massive cartons. The 3D printed toy never saw the inside of a factory injection molding machine or travelled around the world in a carbon spewing network of shipping containers, trucks and planes. Its storefront was a simple web site that didn't need to be heated or lit.

At first glance, it's tempting to heap eco-savior status onto the 3D printing process. But consider the fact that both toys were made of non-biodegradable plastic. Per pound of manufactured product, a 3D printer consumes more than 10 times as much electricity as an injection molding machine.[2] Despite its bad reputation, an injection molding machine is actually very clean and frugal, leaving behind little waste byproduct as it pushes plastic pellets into shape. Finally, a distribution network built on large numbers of small shipments to different locations isn't ecologically efficient. All of this adds up to the fact that if 3D printed manufacturing were merely scaled up to global proportions, there would be nothing green about it.

## Greener manufacturing

The promise of cleaner manufacturing lies in fully exploiting the capabilities unique to 3D printed manufacturing. 3D printing technologies have the potential to disrupt mass manufacturing in the following ways. First, 3D printers can fabricate products whose shape is optimized for its application or environment. Second, storing ready-to-print design files, or digital inventories, is more eco-friendly than maintaining environmentally costly warehouses full of physical inventory. Third, someday distributed 3D printed manufacturing could enable companies to make products locally, near their customers. Finally, 3D printing technologies have untapped potential to work with recycled or earth-friendly printing materials.

### Atkins, a low carb diet for manufacturing

Low carb used to mean a diet restricted in bread, pasta and potatoes. In manufacturing, low carb means "low carbon," a more energy efficient approach to design and production. To researchers at the University of Nottingham (formerly at Loughborough University), low-carb manufacturing means reducing the carbon footprint of the entire product lifecycle, from design, to production, to part assembly, distribution and finally, disposal.

"Current products are generally wasteful in all aspects, from design and manufacture to the final distribution to the consumer," explained Richard Hague, a professor at the University of Nottingham. "This is mainly a consequence of conventional processes that restrict our current design, manufacture and supply chains."[3] Richard and several colleagues conducted an in-depth study to compare the carbon footprint left behind by 3D printing and traditional manufacturing. They called the product the "Atkins Feasibility Study" (a nod to the famous low-carb Atkins diet).

The purpose of the Atkins Feasibility Study was to assess whether 3D printers could shrink manufacturing's carbon footprint. Atkins researchers measured the environmental impact of every facet of the manufacturing process: energy consumption, manufacturing waste by-product and transportation networks. To complete their holistic assessment, Atkins researchers calculated whether superior design and optimized product shape (advantages unique to 3D printed production) could provide environmental benefits later in the course of a product's lifecycle. For example, optimized 3D printed products could be lighter in weight, have superior performance or greater durability.

The results were mixed. On average, compared to traditional manufacturing machines, 3D printers that used polymer-based materials consumed more than 10 times the amount of electricity to make a part of the same weight. Industrial-scale 3D printers that used lasers (or heat) to solidify powdered polymers generated an estimated 65 percent more leftover plastic waste material than did the injection molding process. Some of the printers analyzed used a category of plastic called thermoset plastic that's not recyclable since it tends to lose its material properties if it's reheated or reused. These findings indicate that despite the precision of the 3D printing process, not all 3D printing is a wasteless manufacturing process.

The Atkins study discovered that the manufacturing process for printing printed plastic objects that had lots of large internal hollows was particularly wasteful. Hollow objects need more support material which generates more leftover excess plastic powder. While some excess support material can be recycled, Atkins researchers found that on average, 40 percent of excess raw plastic powder was re-usable in later print jobs while 60 percent typically got dumped into the landfill. The good news is that water-soluble support materials are becoming increasingly popular.

Atkins researchers discovered that printing plastic does have some environmental benefits when compared to the injection molding process, namely the cooling process. Thanks to their slow production process, most of the time 3D printed parts aren't extremely hot after they're made. In the injection molding process, when plastic nurdles are aggressively pressed into a mold they become very hot and require coolants. Frequently, to pry plastic out of an injection mold, factories use toxic chemicals called "release agents."

In contrast to printing plastic, 3D printing metal enjoyed several advantages over traditional metal manufacturing techniques. The Atkins Study found that nearly 100 percent of leftover metal powder from a print job could be re-used. In contrast, traditional metal manufacturing (grinding, machining or molding) processes are more environmentally wasteful. Some metal manufacturing methods leave 90 percent of the raw metal behind in waste byproduct. For example, it can take up to 15 kilograms of raw metal to make a 1 kilogram airplane part.[4]

Since their goal was to study the carbon footprint generated over an entire product lifecycle, researchers studied the downstream impact of 3D printing on the global supply chain. Manufacturing could be greened if companies used digital inventory and local, just-in-time production—a de-centralized manufacturing model that 3D printing is ideally suited for. The Atkins Study concluded that "The application of AM for suitable parts and components, especially those that are of low volume but high value, can result in a significant reduction in stock costs and inventory levels."

One of the most promising (and so far unexplored) environmental benefits of 3D printed manufacturing was subtle: design optimization. The Atkins study said that with 3D printing, traditional criteria for design-to-manufacturing "can be ignored and designers can design what they want or need rather than what the manufacturing system is capable of producing." High-performing parts can help shrink manufacturing's carbon footprint in several ways.

## High-performing printed parts

Computers are great problem solvers. Computer-generated designs create a new breed of products. Reducing weight is an obvious way to shrink a product's carbon footprint. For example, for every kilogram that's shaved off the weight of an airplane the plane will burn approximately 600 fewer liters of fuel per year.[5]

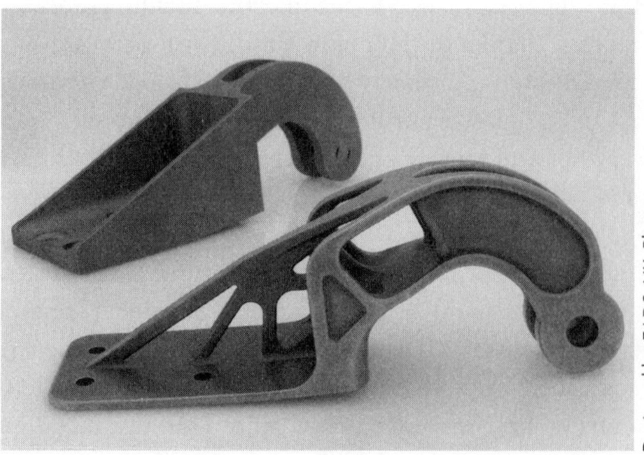

Designed by EADS, UK. Photo courtesy www.paulmcmullin.com

**This metal airplane part was designed by a computer program and then 3D printed in metal. The one in the back is the old version; the one in the front is optimized to weigh less while retaining its strength and other key properties.**

Product optimization takes many forms. A cleverly designed part can last longer or save energy thanks to the fact it's specially designed for its environment. 3D printed custom engine parts, for example, can be designed to carry larger volumes of cooling air or to bear more weight.

Another way to optimize a product is by making it in fewer pieces, or even in a single piece. A general rule of thumb in manufacturing is the more parts in a product, the more resources it takes to make. The more parts that need to be assembled, the longer the supply chain and the larger the inventory.

Thanks to their unique fabrication process, 3D printers can make objects in a single "print job." If future manufacturers could print parts whose designs were optimized for less assembly, the result would be less environmentally costly overhead. A streamlined manufacturing process would involve shipping or fastening together fewer separate parts.

Aerospace manufacturer Boeing found it could 3D print a duct for a fighter jet that in the past was made of 20 separate parts. Once the duct was printed as a single, already assembled piece. Boeing found it could streamline its inventory. Storing the design files and printing parts on demand (rather than storing and tracking an inventory of physical parts) consumed less inventory storage space and reduced administrative overhead.

## Cutting links from the supply chain

Many people, weaned on images of toxin-spewing factories, don't realize that perhaps more devastating than factory pollution is the slow burn of fossil fuels consumed by supply chains. The process of moving materials and parts around the world generates large amounts of pollution. Wal-Mart estimates that about 80 percent of its corporate carbon footprint is generated by its vast and global network of suppliers.

Global supply chains move raw materials to the factory, then to the assembly line, and finally, to the last stop, the consumer. All of us rely on the flow of global supply chains. Nearly every mass produced object we live with, purchase, consume and throw away—from the most humble plastic toy to the medical device that saves lives in surgery—is the product of a long and winding supply chain. Supply chains have a huge carbon footprint because of fuel emissions from industrial armies of trucks, planes, and ships that move things from place to place.

Warehouses that hold unsold and unused inventory consume electricity for heating, cooling, and lighting. Replacing physical inventory with digital inventory would green the supply chain. Physical inventory not only needs to be transported, it also takes up a lot of shelf space while it waits. In contrast, a digital inventory—or design files for a 3D printed machine part—is cheap and easy to store and transport.

3D printing technologies could help clean up the manufacturing process if their unique capabilities are put to use. A looming challenge, however, is the end of a product's lifecycle. Think back to the hypothetical example of the two plastic toys earlier in this chapter. Both were made using standard industrial plastics.

If both toys were thrown away, would they end up in the same place? Sadly, the answer is probably yes. The problem lies in plastics. 3D printers use essentially the same sort of commercial plastic as injection molding machines do. Like a younger sibling eating the same food as his older siblings, since 3D printers grew up on the factory floor, they have retained an appetite for the same raw materials used in mass manufacturing.

## 3D printing a more beautiful landfill

Exotic 3D printing materials get a lot of media attention, for example chocolate or gels containing living cells. Other printing materials such as metal, ceramic, and glass are slowly finding their way into some industrial uses. Yet according to market data tracking the sale of printing materials, plastic still reigns supreme.

Lurking near a 3D printer are bound to be buckets full of plastic powder or spools the size of a hubcap wrapped in brightly colored plastic strands. Powdered forms of nylon are popular. Other popular printing plastics in powder form are polypropylene (same as the plastic that makes up your yogurt containers) and polyethylene (found in trash bags).

MakerBot test-drives each new printer before shipping it to its buyer. When I visited its former production facility in 2012, in Brooklyn, New York, rows of brightly colored spools of ABS plastic were suspended over the printers, lending the scene a cheerful appearance, reminiscent of a brand new box of crayons. Similar to MakerBot's printers, most low-cost 3D printers popular with consumers use a type of plastic called ABS that's found in LEGOs, white-water plastic canoes and hard-bodied suitcases.

There are a few reasons that plastic remains the queen of 3D printed materials. First, plastic is cheap and easy to work with. Second, plastic has a long and successful track record as a raw material for mass manufactured objects, ranging from simple bottles to elaborate and expensive boat hulls.

Sometimes the word "plastic" is used disparagingly to describe a person or object that's not genuine. In fact, plastic is very real. It's so real that it's a major (and rapidly increasing) source of pollution to our planet. Twenty-five times as much plastic was produced in 2000 compared to 1960.[6]

Journalist Susan Freinkel traces the shift in our society from an initial infatuation with plastic to our current relationship, a highly dependent relationship she describes as a "toxic love affair." When plastics first were developed nearly 150 years ago, they were heralded as a democratizing new material that would save tortoises and elephants from extinction.[7] After World War II, plastics came into widespread public use. According to the American Chemistry Council, plastic became the world's most widely used material in 1976.

Most plastics today are made from fossil fuels, both natural gas and petroleum. One reason plastic costs so little to make is because it's made from waste byproducts produced when fossil fuels are extracted from the environment.[8] In a demonstration of how just a few molecules can completely change an object's characteristics, plastic contains a high amount of carbon and hydrogen, the elements that characterize living objects.

A plastic object's ability to endure rough handling and challenging conditions is what makes plastic so useful. Yet the same tenacity and endurance also make plastic objects environmental malingerers, a rapidly growing zombie army of waste that refuses to biodegrade, no matter how long or far it travels. From the garbage can, plastic objects travel to the landfill. Or they end up washing into the ocean.

In 1997, during a sailing trip to a remote part of the Northern Pacific Ocean, Charles Moore was shocked and dismayed to find his boat surrounded by a vast floating tangle of discarded plastic "confetti." Moore dubbed this glut of floating plastic particles—a region whose size is estimated to be slightly smaller than that of the state of Texas—"The Great Pacific Garbage Patch."

After this wake-up call, Moore wrote about his experience in his book *Plastic Ocean*.[9] He made it his life's work to begin collecting and analyzing the plastic debris littering our ocean waters. Over the years, Moore and his crew have regularly traveled to the Great Pacific Garbage Patch to pull out and catalog tons of floating plastic waste, most of it in the form of tiny particles.

Most plastics that wind up floating in our oceans won't ever biodegrade—they're here to stay. Over time, the ocean's movements break floating plastic garbage into tiny pieces—the plastic "confetti" described by Moore—that are invisible in satellite photos but deadly to the marine ecosystem. The gigantic floating plastic garbage islands in our oceans continue to grow each year, disrupting marine life, choking birds and seals and leaching toxic byproducts into the ecosystem.

Moore reports that the amount of plastic confetti captured by fishing nets outweighs that of zooplankton, the ocean's food base, by a factor of six to one. In addition to tiny plastic fragments, his nets also pull in more recognizable everyday plastic objects such as disposable lighters, plastic fishing nets, plastic handles, children's toys and of course, plastic bottles.

Today, the production of 3D printed plastic parts is miniscule in volume compared to the tidal wave of plastic goods mass produced in factories. If placed next to the Great Pacific Garbage Patch, the amount of 3D printed plastic garbage would be tiny, a child's shoe next to a football field. Yet, like any other plastic object, whether it be custom-made or mass manufactured, most 3D printed products that reach the end of their lifecycle will be tossed into the garbage can.

3D printing pioneer and outspoken visionary Joris Peels points out that if we continue to consume and discard products at our current global rate of consumption, we will choke to death in our own waste. In a blog article entitled "3D Printing vs. Mass Production: A More Beautiful Landfill," he wrote, "My fear is that eventually mass production could lead to mass extinction . . . I really believe we're headed to the path to extinction . . . Like the Easter Islanders we're also going to cut down the last tree."[10]

Plastic is environmentally devastating. Yet plastic has also been a great democratic equalizer, enabling nearly everyone to own household items that used to be reserved for the rich. Plastic objects are found in nearly every aspect of our lives, from mundane plastic toys to lifesaving lightweight plastic tubes used to give blood transfusions.

Plastic car parts are lighter than metal in engines and car interiors; savings in weight have environmental benefits. Plastic packaging saves lives by preserving food. Plastics make jobs: The plastics industry is one of the largest employers in the United States. The packaging industry, a cousin to the plastics industry, is even larger; worldwide, the packaging industry ranks third to only food and energy.

For better or for worse, our economy revolves around plastic goods. By giving regular people the power to make things from plastic at home, 3D printing opens up yet another new channel of plastic manufacturing. To become a greener form of manufacturing, 3D printing technologies need to embrace new, ecofriendly raw materials.

## Printing recycled milk jugs and dirt

To learn more about green materials, I spoke with Mark Ganter, an engineering professor at the University of Washington in Seattle. Mark explained that "Most of the printing materials on the market are not really very Earth friendly." Mark and fellow professor Duane Storti run the Solheim Lab, a hub for radical 3D printing at the University of Washington. Years ago, Mark and Duane Storti both sacrificed a month's salary to purchase a 3D printer for their lab, one of the first on campus. They haven't looked back since.

One of the lab's major research areas is biodegradable 3D printing, including developing and testing ecofriendly and recycled materials. Students in the Solheim lab 3D print in "everything from sugars, to wood, to crab shells," said Mark. One student 3D printed an experimental mixture that involved adding blood as the adhesive, because of its good clotting capacity.

Mark explained that a 3D printer can fabricate "basically anything that can be powdered to an appropriate particle size, and I mean anything." The key is to grind a powder down to the right fine-ness and texture to allow it to be spread in paper-thin layers. Once a material is ground into powder, an appropriate adhesive must be found. This adhesive powder is mixed with the material powder and the mixture is placed into the print bed, and spread into a paper-thin layer of mixture. The print head then lays down a solvent to activate the adhesive material to hold the powder together to form layers.

**Objects printed from terra cotta**

One of the Lab's bolder experiments in green printing finished second in a local annual boat race in Seattle, a 3D printed boat made from recycled milk jugs. The Lab's student organization, Washington Open Objects Fabricators Club (also known as "WOOF"), designed and printed a boat and entered it into the Milk Carton Derby, a high-profile boat race that's part of Seattle's annual Seafair Festival. Derby race rules are strict. Only the following cartons may be used to provide flotation: half gallon and one gallon plastic and paper jugs that held milk or juice.

The WOOF team began work on their boat weeks before the race, in a dumpster. Students went dumpster diving and hauled nearly 40 pounds of plastic jugs back to the lab. They ground the plastic jugs into fine powder and hacked a 4-by-8-foot plasma cutter with a homemade extruder. To power the printer's plastic extruder, undergraduate Matthew Rogge ripped out the windshield wiper motor from his Subaru. Over the course of two months and several failed test runs, students learned that printed milk jug powder, upon printing, is prone to shrink in size about 2 percent. After some tinkering with the design, the WOOF team spent two days printing a boat that could support 150 pounds and cut through the water "like a canoeyak," according to Mark.

Image courtesy of Brandon Bowman, Washington Open Objects Fabricators (WOOF) club, and Mark Ganter and Duane Storti, Solheim Lab, University of Washington

**A 3D printed plastic boat made by grinding up used milk jugs**

The day of the Derby when the team showed up with the printed boat, "the boat required some explaining," Mark Ganter said. After some discussion with the Derby race committee, it was agreed that the printed boat could race in the age 14 and up group. After getting last-minute permission from Derby judges, five minutes before the race, students Matt Rogge and Adam Commons loaded the boat into the water. Matt Rogge paddled his way to a second place finish in the 14+ age group.

The 3D printed boat was an environmentally exciting project because it proved that recycled milk jugs can be re-used and 3D printed into substantial and usable objects. Plastic milk jugs are made of HDPE (high-density polyethylene) plastic, a widely used and recyclable type of petroleum thermoplastic. Most plastic doesn't get recycled, however. According to the U.S. Environmental Protection Agency, across all types of plastics, recycling rates average about 13 percent, much lower than that for garbage made of glass, steel, aluminum, and paper.[11]

Plant-based plastics may provide a greener alternative to petroleum-based plastic. When I asked Mark about available greener printing plastics, he described PLA as "a great material for FDM 3D printing. Soy plastics are also a good choice." PLA is a commonly used corn-based thermoplastic for 3D printing that is a water-soluble thermoplastic. PLA can be used for support material and, since it's water soluble, can be rinsed off with water (not solvents) and reused.

Another good way to make plastic printing more ecofriendly would be to recycle and re-sell used ABS plastic printing filament. "I hope someone starts selling recycled plastic filament soon," Mark said. "Our lab and others have been experimenting with this idea. What if every home could turn its household plastic waste into usable plastic filament for 3D printing? Wow!"

Maybe the first re-cycled plastic filament will soon be commercially available. A student at Vermont Technical College, Tyler McNaney, raised $10,000 on Kickstarter to build a device that grinds up and re-melts discarded 3D printed plastic objects into printing filament. Tyler named this recycling device the Filabot, described on its website as "user friendly, but … also environmentally friendly. Filabot will bring the real power of sustainability to 3D printing."

Recycling plastic is a good start. But more is needed. "To make 3D printing a technology that can advance life, we need to find ways to print in waste product, food byproducts, recycled glass, sand, even dirt," said Mark. He plans to do more research to explore whether it's possible to use powdered food materials as a sustainable raw material for 3D printing. "Rice flour is available almost worldwide (and it prints quite well, thank you)," he laughed. "And food byproducts, like corn husk or wheat chaff, cost very little."

"Green 3D printing seems like an amazing direction to take. Re-purposing waste into useable objects might be a fairytale. But it's an idea worth taking a good hard look at to see if we can make it happen," Mark concluded.

## From "planned obsolescence" to "exuberant waste"

One environmental hazard of 3D printing has nothing to do with the manufacturing process or raw materials. The risk is a new mindset. 3D printing gives people the power to design and make whatever physical object they can dream up. But like the old saying goes, "Nothing in life is free." Giving people tools of production also introduces the temptation to become wasteful.

When they were introduced a few decades ago, low-cost laser printers did not lead immediately to the paperless office; instead cheap and widely available printing technologies encouraged people to print casually and wastefully. Similarly, ubiquitous 3D printers might lead to the production of more junk, making it easier for people to fabricate objects in a wasteful spirit without realizing the downstream costs of their actions.

Engineers, seamstresses, and probably even surgeons are advised to measure twice and cut once. When production is costly or risks are high, a designer or engineer will measure and plan again and again to make sure physical production will go as planned. Easy-to-use design software and a nearby 3D printer make it easy to not heed this ecofriendly approach. Unfortunately, 3D printing induces a spirit of "irrational fabrication" into some susceptible souls.

I've seen the effects of irrational fabrication first hand. One morning I came into the lab and discovered two dozen or so misshapen plastic objects of similar shape lying in casual disarray on the table next to the lab's 3D printer. It turned out that the creator of this trashcan full of new plastic garbage was one of my students, who for a class assignment had spent the night in the lab, feverishly 3D printing faulty prototype after faulty prototype. Like a frustrated author ripping out sheet after sheet of paper, this student printed out a design, slightly adjusted its dimensions in the design file, then printed it out again, hoping the next one would turn out better than the previous one.

Software engineers debug half-baked bits of code by compiling it and testing in digital form. However, compiling and re-compiling bad code doesn't burn up precious raw material. It burns up time and a developer's patience, but doesn't fill the garbage can with badly shaped plastic objects.

Now that 3D printers have sped up the prototyping process, the environmentally unfriendly physical debugging process has become a real option, at least for small objects. When people 3D print half-baked design ideas rather than testing them first in a computer simulator or measuring twice or for the third time, they're applying what some might describe as a debugging mentality rather than a design mentality to their product design process. Until recently, a debugging mentality in the physical world was costly and time consuming. 3D printers make it easy for people to test their design files by printing them, something they would not dream of doing had they had to fabricate the parts by carving them by hand or using mass-manufacturing techniques.

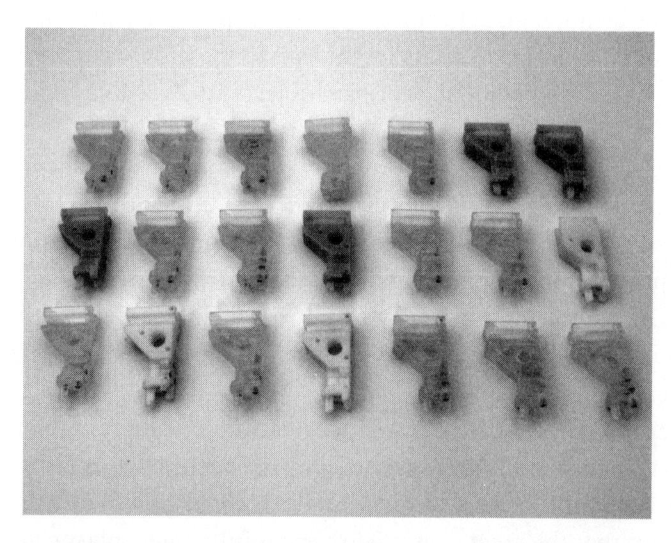

**Failed prints. Are we switching to design by trial-and-error?**

In a sign of things to come, if you visit any shop or lab that offers students or engineers unlimited access to a 3D printer, any flat surface will be littered with failed 3D printing experiments. There's a longer term upside to physical debugging, however, in situations where complicated designs can't be properly examined using simulation software. As designs become more complex, it's nearly impossible for even a skilled engineer to stare at it on a computer screen and predict whether it will work in real life. Good computer modeling software can help simulate a designed object's physical characteristics. But there's still no substitute for the physical thing.

Quickly fabricated 3D printed prototypes can save waste on mass manufacturing assembly lines. By making a physical version of a digital design file, engineers and designers get one last chance to make sure their product plans will work. Design problems and aesthetic shortcomings can be corrected before the mass manufacturing machines kick into high gear, burning up precious energy and churning out thousands of copies of a mistake.

3D printing technologies offer both promise and peril for the environment. It would be disappointing if 20 years from now, Charles Moore were to fish out 3D printed plastic toys, custom shower curtain rings and other plastic tchotchkes from the Pacific Garbage Patch. In an ideal world, all manufacturing would be as clean and innovative as the desert-based Solar Sintering process. 3D printing won't be an innately green manufacturing technology unless we actively seek to make one. If we can tap into 3D printing's unique capabilities and invent greener printing materials, we will reap environmental benefits in the form of shorter supply chains and a new generation of optimized products.

# 12 Ownership, safety, and new legal frontiers

Law can be one person's protection and another person's hindrance. Especially in laws that attempt to define philosophical issues such as safety or ownership. I experienced this first-hand while teaching a special summer seminar for high school students interested in design and engineering. The week-long curriculum taught product design principles using simple design software and our lab's 3D printer. The plan was that each student would design a product, name it, describe its commercial value and finally, post it for sale on Shapeways.

At the end of the week, students had created a rich variety of product designs. On the last day of class, I demonstrated the final step: Each student would upload their product design file to Shapeways where it would be sold online.

One student, who at the start of class had cheerfully showed his friend how to jailbreak his iPhone so he could download free music, indignantly raised his hand: "But if I upload my product design file onto the Internet, what if somebody just downloads it and makes a bunch of free copies without paying me for them?"

Another student said that maybe she shouldn't upload her product design after all. She wasn't sure whether the ornate iPhone holder for bicycle handlebars was secure enough to prevent someone's phone from slipping and shattering on the street. Another student explained that he didn't care so much if someone used the files for the 3D printed puzzle he had designed. What would really bother him, he explained, would be if someone took his design and then claimed credit for it.

It took this class of high school students just a few minutes to express the legal challenges that lie ahead. Like the computing industry a few decades ago, the marketplace surrounding 3D printers and related services is still in its infancy. It's an open frontier, a relatively unpopulated space whose commercial activity is still too modest to have come to the attention of lawmakers, corporate litigators and criminals.

The downstream impact of emerging, game-changing technologies is difficult to predict. Look at your personal computer, so innocently sitting there on your desk. In just a decade or so, computing has shaken up our legal system at its foundations.

Local sales tax or value added tax (VAT), once a simple and straightforward issue, has become a complicated accounting question for Internet-based commercial transactions. The crime of "stalking" has taken on a whole new meaning online. Consumer privacy laws are stretched by companies that shamelessly track their customer's internet searching, browsing, and buying habits. Organized crime is no longer a centralized, local phenomenon as cybercriminals commit fraud and conduct espionage from remote and hidden locations.

When computers were a costly industrial tool doing their work in the backrooms of data centers, they didn't stir up a whole lot of new legal challenges. However, when computing power reached a critical mass of the population, people and businesses quickly discovered that existing laws and regulations were woefully inadequate. Core legal definitions of ownership, location and format had to be redefined.

3D printing, like any industry that experiences rapid technological leaps forward, will also experience new legal challenges and novel forms of consumer safety and criminal activity. Law changes slowly. But technology doesn't wait.

## Printing weapons, drugs, and shoddy products

Being a currency counterfeiter used to be a skilled profession. In a recent public service bulletin, the U.S. Secret Service noted that today's counterfeiters are a new breed. Modern counterfeiters need only "basic computer training and skills afforded by trial and error, and public education."

Before 1995 fewer than one percent of fake bills were made using computers and laser printers. Just 5 years later, in 2000, just under half of false bills were designed online and then printed out on a high-end color printer.[1] Computer design software, color laser printers, and toner technology have democratized currency counterfeiting. Traditional offset printing, the old method, demanded a skilled counterfeiter with years of experience.

Counterfeit money causes economic harm. However, 3D printed counterfeit goods and unauthorized weapons could cause bodily harm. One of the first moral skirmishes in the 3D printing community took place in 2012 on a file-sharing site called Thingiverse.com.

A Thingiverse.com user uploaded a design file for a 3D printed rifle part that could be made out of plastic on a consumer-scale 3D printer. The uploaded design file enabled people to skip a critical part of the control process of getting a license to carry a gun. This particular gun part was the only piece of this rifle model that required that its users undergo a background check. In other words, by 3D printing this gun part, a person could sidestep gun control laws.

The law didn't get involved in the Thingiverse situation. Instead, the 3D printing community's response was to ponder the issue. Eventually, after much discussion, the decision was made to ask that the gun part's designer remove his file from the website.

Image courtesy of Michael Guslick

**A .22-caliber gun manufactured partly with 3D-printed plastics**

A few months after that first case, a gunsmith by the online name of Haveblue fabricated a functioning firearm made partly out of 3D printed plastic parts. The .22 caliber gun was a hybrid of 3D printed and commercially purchased parts. The main body of the gun was plastic but the chamber that held the bullets was metal.

Haveblue reported online that the printed plastic gun parts were robust enough that he was able to fire 200 rounds. No special equipment was needed. The gun parts were printed on a fairly old Stratasys machine using normal commercial-grade resin. The gun's creator estimated that it cost about $30 to buy the resin to print the gun parts.[2]

It was always possible to make your own gun, but 3D printers are the ideal tool for tech-savvy criminals. A 3D printer creates objects shaped in ways that were once impossible to make by hand or by machine, for example, a gun that looks like shoe or a hairbrush. 3D printers are small and portable. They can make one custom object after another, in stealth, no factories, coordination or unnecessary exposure needed.

Counterfeiting and crime—like mainstream manufacturing and design industries—are made more efficient and innovative by new technologies. Illegal arms trading is big business. So is drug trafficking.

## Designer drugs

Maybe someday you will tell your grandchildren about the old days when people had to have a doctor prescribe them certain drugs and you had to get security clearance to buy dangerous chemicals. "Really?" your enthralled listeners would exclaim, their eyes wide.

"Yes," you would nod. "That was way back before the days of DIY chemical factories."

In an article by writer Nikki Olson, Lee Cronin, a professor at the University of Glasgow, described his most recent project, a mini chemistry lab built on a low-cost 3D printer. Lee Cronin and his team used an open-source printer (Fab@home) to print a mini chemical factory, or what they describe as "reactionware."

Reactionware is a polymer gel whose internal structures contain vessels that give it a special shape that enables chemical reactions to take place. Like a glass test tube that remains aloof from the chemical action that takes place inside of it, reactionware offers a neutral environment that enables chemical reactions to take place undisturbed. Cronin and team tapped into the precision and digital control of 3D printing to fabricate a custom gel containing intricate vessels that can catalyze chemical reactions at a much lower cost.

The notion of running chemical reactions inside a special container isn't new. Vessels, or containers, that contribute and catalyze chemical reactions have been a necessary part in any large commercial laboratories for years now. What's disruptive is the ability to catalyze sophisticated chemical reactions using low-cost tools that are readily available.

Combinations of chemicals and materials that were once difficult to bring together open up a new design space that will allow chemists to explore infinitely more types of new chemical compounds. This means fast exploration and innovation, but could also mean new unregulated substances, or new recreational drugs.

In the abstract of his breakthrough paper on this project, Lee Cronin describes its potential: "Three-dimensional (3D) printing has the potential to transform science and technology by creating bespoke, low-cost appliances that previously required dedicated facilities to make."[3] In other words, the central control exercised by the drug and chemical industries will be challenged by low-cost tools of design and production.

The more powerful the technology, the more grotesque and far-reaching its potential abuses. So-called "do-it-yourself" (DIY) drug production breaks down the relatively uniform and consistent recipes for making drugs and compounds. Having people design and produce their own drugs and powerful chemicals could present a regulatory nightmare.

Today's War on Drugs in the U.S. has already failed miserably. Prisons are stuffed full of non-violent offenders and precious tax dollars are spent on arresting drug users rather than on lower-cost and more effective drug rehabilitation programs. Deaths from prescription drug overdoses have soared in developed nations. Imagine the devastation if people could fabricate a batch of their favored mood-altering chemical concoction at home by printing the necessary reactionware.

Mini chemical factories would introduce another risk: unrecognized chemical substances. Most drugs and household substances are classified into recognizable categories. Therefore, in the case of accidental poisoning or a drug overdose, medical professionals have a standard framework that helps them understand the chemical's effect on the body and how to treat it. If people were to print up designer drugs or solutions, the medical profession would have a much harder time figuring out what, exactly, the person ingested and what the antidote should be.

However, low-cost chemical production has the potential to help people as well. Cronin said, "We could see 3D printers reach into homes and become fabricators of domestic items, including medications. Perhaps with the introduction of carefully-controlled software 'apps,' similar to the ones available from Apple, we could see consumers have access to a personal drug designer they could use at home to create the medication they need."[4]

## Consumer safety

Another risk introduced by 3D printing technology will be consumer safety. Consumer safety is something that most of us in industrialized nations take for granted. What's easy to forget is how many legal and regulatory safeguards have been put in place over the years to make sure that we don't get hurt by the products we buy.

Nobody has been hurt yet by a malfunctioning 3D printed machine part or toy or by a 3D printer itself (at least not seriously). One of the most difficult legal aspects in consumer safety cases isn't capturing criminals. It's figuring out who's legally at fault when failure happens.

Tort law is less appetizing than its name at first suggests. A torte is a rich cream filled cake. Tort law, in contrast, untangles blame (or innocence) of involved parties when a product malfunctions and someone is accidentally hurt or killed.

Under tort law, a party's responsibility ends if that party can demonstrate it made reasonable efforts to head off and minimize the risk of product failure under reasonable use scenarios. For example, a tire manufacturer would be held responsible if its tires blew up every time a vehicle drove over 80 mph; even though this is an illegal driving speed, it is a likely use scenario. However, that same tire manufacturer would not be held responsible if the tire blew up after a buyer of the tires rammed his car into a curb.

Here's a scenario that may someday be played out in the courtroom. Which party is at fault in this hypothetical case? A car enthusiast with good intentions 3D prints a custom-steering wheel in her basement workshop. Needless to say, this printed steering wheel didn't undergo quality and safety control procedures. The design file was purchased from a popular website that sells flashy novel car add-ons such as novelty air fresheners and gear shift handles.

The amateur car enthusiast sells the printed steering wheel online. Its buyer installs the wheel in his car. A few weeks later, he discovers too late that the custom printed steering wheel disengages if turned sharply to the left at high speeds. Its buyer dies in a tragic car crash.

Imagine you're the attorney defending the family of the amateur car enthusiast who died in the car crash. Where would you assign fault? On the person who made the faulty design file? The person who 3D printed and sold it? The website that advertised the part? Perhaps the manufacturer of the car the faulty wheel was installed onto, the printer manufacturer, or the material supplier?

Standards can help set clear boundaries of responsibilities. In the early days of steam engines, boilers used to blow up, frequently causing injury and damage. It was insurance companies that insisted on delineating responsibilities by setting clear standards of production. A set of criteria were eventually created that specified the minimum requirement to certify a boiler for a certain operating steam pressure, such as material thickness, safety margins and pressure release valves. A boiler that did not meet the standard would likely not be insured.

Similar standards will develop for 3D printing to help create boundaries of responsibility. Printer manufacturers will strive to get their printers certified, and material producers will try to meet or exceed minimum material performance standards. It's possible that in the absence of clear and standard boundaries of responsibilities, manufacturers of complex 3D printed products will have no recourse but to disclaim all responsibility altogether—a situation you will be familiar with if you ever read the end user license agreement you hastily agree to every time you install new software. Most software is sold "as is, without any warranty of fitness for any particular purpose."

Criminals will quickly learn to apply 3D printing technology to improve their illegal wares and services. All of us face new risks if we have the bad luck to purchase a shoddily made or counterfeit 3D printed machine part that fails at a critical moment. 3D printed weapons and new chemicals could be devastating if they fall into malevolent hands. Far in the future, maybe the black market for organ donation will shift into a black market for unregulated and bioprinted body parts.

In reality, these dramatic concerns will likely not affect most of our daily lives that much. I suspect that the legal challenge that most of us will run into will be that of navigating out-of-date intellectual property laws.

## Rip, mix, and burn physical things

In the Middle Ages people fought to gain control over tracts of land. Some economists trace the origins of modern capitalism back to medieval times, when communally-owned lands were partitioned into privately owned units. Today, companies fight to gain control over units of intellectual property. Commercially valuable ideas have become the new land that forms the foundations of our modern economy.

When digital media exploded into consumer markets, the infamous Napster case marked an inflection point, the formal declaration of war between music consumers and the entertainment industries. The world of 3D printing has not yet faced its own large-scale "Napster moment." People speculate that big aggressive companies known for fiercely guarding their intellectual property—the toy companies, software companies, and media conglomerates—don't yet feel that they're losing money because of unauthorized 3D printed copies of their products.

Today's "wait and see" attitude may change when 3D printing technologies gain enough critical mass and commercial momentum to take a bite out of sales. In the words of journalist Peter Hanna, "If the current 3D printing free-for-all sounds too good to last, it is. The community today is small and has avoided, either by chance or design, stepping on any really big toes." [5]

Hardened lawbreakers are one thing. However, enthusiastic consumers and small business owners are another. Most people and small businesses would prefer a clear set of legal guidelines around 3D printing.

Let's explore a hypothetical scenario. A small business or an individual hobbyist innocently 3D prints a small plastic figurine that's an exact replica of an iconic and copyright protected cartoon character. The toy's creator puts the figurine onto his website and sells it for $20.

After a few weeks, lawyers at a global media and toy corporation see the printed replica for sale online and send a "cease and desist" letter. The charge? That by making an unauthorized copy of a copyrighted toy without asking permission and paying royalties to the global company, the creator of the 3D printed toy is in violation of Big Corp's copyright. The letter suggests that the toy's creator either immediately remove the printed figurine from the website or negotiate a licensing deal.

At this point, if the toy's creator is a person with innocent intentions who just wasn't aware of copyright law, he would sadly and quickly withdraw the 3D printed toy from their website. Or, perhaps the toy's creator just doesn't have a strong enough stomach or big enough bank account to hire their own attorney and dive into legal battle with Big Corp. End of story.

What if, however, the situation played out another way? What if the toy's maker, upon receiving the cease and desist letter from Big Corp, decided that he's not actually in violation of their copyright. Perhaps he would compromise. He would agree to quit selling the original offending plastic printed toy. Instead, he would design and sell a new set of printed figures: little plastic figurines that boast the heads of various presidents from countries all over the globe. The heads of different presidents will be attached to the bodies of familiar Big Corp characters. Such design modifications would be quick work with design software and a 3D printer.

The saga continues. Big Corp is not amused and sends the intrepid toymaker yet another cease and desist letter. At this point, most people or small businesses would probably just give up. Most people or businesses can't afford to sue a giant global corporation and live to tell the tale, particularly if they lose.

The hypothetical saga continues. A philanthropist with a passion for protecting civil liberties sets up a legal defense fund to address exactly these sorts of issues. This philanthropist, upon hearing of the case, would step in and offer to pay legal bills. The person selling the printed figurine is an idealist. He strongly believes that copyright laws need to be re-zoned in order to provide a fair and workable framework for 3D printing. He steps up to the challenge and agrees to push the legal battle against Big Corp into high gear.

What would happen next? We don't know yet. Big Corp would likely weigh potential legal costs against the relatively tiny dent in sales caused by the toymaker's 3D printed figurines. In this case, Big Corp's executives would probably decide that bad press, plus legal costs, would not be worth a little lost revenue. If millions of small businesses and consumers started 3D printing copyrighted toys, however, Big Corp would probably eventually swing into action. We could replay this hypothetical toy scenario a few different ways. If the situation involved a 3D printed patented machine part, a similar legal situation could result.

At the time of this writing, a full blown legal battle over intellectual property infringement hasn't yet occurred. Given the legal precedents set by digital media lawsuits, however, the writing is on the wall. A few quick legal scuffles over intellectual property rights have already taken place.

A youngster in the United Kingdom designed and 3D printed a plastic replica inspired by one of his brother's action figures. Paramount, the company that made the action figures, was not pleased. Their attorneys sent a threatening "take down" letter to the boy, warning him that his printed toy was in violation of the company's copyright. Shaken by the unexpected legal threat, the boy sadly complied.

Intellectual property law is a sword that cuts both ways. Big companies can squash competitors and innovative technologies by intellectual property as their weapon of choice. However, small inventors and artists rely on intellectual property rights so they can receive fair payment from people or companies who financially benefit from their inventions or works of art. The high school student in our summer design and engineering class who was concerned with intellectual property wasn't concerned about being "ripped off" by a big corporation. He was concerned about his work taken by another maker, without credit.

The challenge lies in defining a workable legal framework to make sense of the vast expanses of gray areas and differing perspectives. It's relatively simple to just enforce existing intellectual property laws when the culprits are career pirates or counterfeiters who maliciously and intentionally disrespect other people's intellectual property rights. However, as 3D printing technology reaches the mainstream, simple "bad guys" will be the exception.

## Trademarks

Once when I was in the Caribbean at an open air market, I bought a cheap duffel bag that had a Nike insignia on the side, but with one minor modification: the Nike swoosh was crossed with another swoosh, a smaller vertical one. The reason the swoosh on the Nike-like bag I bought in the market had the additional cross-swoosh worked into the design was to protect the bag's maker against violating Nike's trademark.

Like patent and copyright, trademark is considered an intellectual property right. People lump trademarks in with the notion of a "brand name" or "logo." A trademark is actually more specific than that: it's actually a registered mark, or "trade dress" that signals to consumers that a product is made by a particular manufacturer.

The original purpose of a trademark was to protect consumers. Since then, a trademark has become more of a marketing tool. A highly recognizable trademark can be worth billions of dollars.

Trademarks, like patents, must be obtained in a central government office. You can't just register a cool-looking logo as a trademark. The product you're marking must be commercially for sale. If 5 years pass (this varies according to country) and you're not selling your trademarked good commercially, your trademark is considered abandoned and anybody can use it.

## Copyrights

Patents protect inventions that have some kind of useful value. Trademarks must be commercially active. In contrast, copyright applies to creative works.

The key to copyright is that the work of art is expressive, not utilitarian, nor useful. As explained by attorney and 3D printing expert Michael Weinberg, "Objects that do things can't be protected by copyright. Useful is drawn broadly. Clothing is a useful object. You can copyright a pattern on a piece of clothing, but you can't copyright the cut."[6]

Like a patent, a copyright gives a creator of original work exclusive rights to it for a specific period to time (the time period varies by country but a typical time span is the life of the creator plus an additional 70 years.) People or companies must ask permission of the copyright owner for the right to reproduce or modify the work, to sell or lend the work, to publicly display the work, or to perform the work in public. People do not need to ask permission if they plan to use an original work for what's called "fair use," for example, in a classroom.

Unlike a patent, a copyright is not granted via a formal application process by a central government agency. Copyright protection appears when a creative work is fixed into some kind of medium. Governments offer formal copyright registration. Though formal registration may be useful downstream to establish priority and ownership should a lawsuit arise, it isn't required.

People who hold copyright can attempt to combat infringement by suing the alleged infringer. However, before kicking off a lawsuit, the copyright holder must first ask the alleged infringer to stop what they're doing, what's called a "take down" notice. If an infringer didn't know she was infringing and quickly takes down the copyrighted work, there will be no trouble. The problem with this scenario is that frequently, the alleged infringer may not actually be infringing, but most people don't want to go to court to find out.

We suspect that one of the thorniest issues in future copyright battles will be resolving what's known as "derivative works." In copyright law, a derivative work is a creative work that incorporates some previously published copyrighted material. Translations are considered a derivative work. Parodies of popular songs are a derivative work.

If you edit the design file of an existing copyrighted object to make a new object, at what point are you creating a derivative work, at what point are you creating a fresh brand new work (meaning you're the creator)? Or, intentionally or not—are you just plain old ripping off somebody else's creative efforts?

If you browse any website that offers 3D printed goods or design files for sale or swap, you will probably encounter a few items that are probably under copyright. The reason big companies that deal in brand name products aren't yet suing small designers and 3D printing enthusiasts isn't goodwill, it's money. 3D printed goods that could arguably violate someone's copyrights are not sold in high enough volumes to trigger the alarm bells at a global toy or consumer goods company.

## Patents

The word "patent" comes from the Latin word *patere,* which means "to lay open." A common misconception about patents is that getting a patent means you keep your idea a secret. It's actually the opposite. When an inventor is issued a patent on an invention, the country's patent office makes publicly available the information necessary to enable another person "skilled at the art" to replicate that invention.

One of the reasons governments set up a centralized patent office was to give inventors incentive to publicly disclose their inventions. According to Wikipedia, a patent is in effect "a limited property right that the government offers to inventors in exchange for their agreement to share the details of their inventions with the public. Like any other property right, it may be sold, licensed, mortgaged, assigned or transferred, given away, or simply abandoned." [7]

There's a significant legal difference between a patent and a trade secret. A trade secret is when an inventor prevents competitors from using the same product idea by simply not telling anyone how the product works. A trade secret is never made public, nor does it get a formal state-issued period of protection, nor does it ever expire.

To be patentable, an invention must be a new, useful, and non-obvious process, machine, article of manufacturing, or composition of matter. In the United States, a patent lasts 20 years from the date that its application was filed with the patent office. Once a patent is issued, it provides its inventor the right to exclude others from practicing the invention. Inventors apply for a patent at a centrally managed government agency, which in the United States is the U.S. Patent Office. An inventor on a patent application can be a single person, a group of people, a single company, or several companies.

The only way to enforce patent rights is in court. If you're an inventor and you own a patent and think another company is using your invention without your agreement, you can sue them for patent infringement. Untangling ideas in court is a notoriously tricky business. For example, a single technology—say a mobile phone—can include thousands of technologies covered by patents owned by hundreds of different inventors.

The challenge is that the act of copying and 3D-printing objects does not fall neatly into any of the above categories. If the product's technology is not covered by an active patent, there's no patent violation. If there is no registered trademark embedded in the 3D objects, there's no trademark violation. And finally, if the physical object has some utility, it is not covered by copyright either.

## Digital rights management

Intellectual property law is difficult to enforce. Particularly when billions of people all over the world are happily making copies of things and sharing them with their friends. Courtroom battles don't scale. Media companies that try to enforce their intellectual property rights in court fight a losing battle. As MIT professor Neal Gershenfeld said, "You can't sue the human race."

Since lawsuits are too expensive, slow and ineffective—particularly against consumers—a common tactic big companies use to defend their market turf are technological safeguards or digital rights management (DRM). For example, Apple applies DRM to its iTunes files to prevent people from making more than a fixed number of copies. When people "jailbreak" their iPhones, the reason they are able to download anything they want without paying royalties is because they've removed the DRM. DRM is why an eBook book can't be copied onto another reader.

Commercially sold design software and 3D printers don't have a formal system of digital rights management . . . yet. In 2012, an arm of patent brokerage company Intellectual Ventures was issued Patent #8,286,236. The patented invention is for digital rights management for manufacturing machines, including 3D printers. In a nutshell, design files would contain digital rights management technology. A 3D printer reading a DRM design file would refuse to print it, similar to the way a software application refuses to work after its product key has expired.

DRM technologies may be a futile attempt to stem the tide. DRM technologies create an ongoing arms race between consumers and companies. As far as 3D printing goes, there's a second complication: patented or copyright objects that are scanned and then 3D printed aren't captured by digital rights management implanted into design files. As pointed out by Simon Bradshaw, Adrian Bowyer and Patrick Haufe, "It is the sharing of (as seen, legitimately) reverse-engineered designs that is the issue, not original design documents." [8]

Perhaps in anticipation of yet another DRM struggle ahead, the Free Software Foundation (FSF) created a certification program called Respects Your Freedom (RYF). The Foundation will certify hardware vendors whose products meet the Foundation's standards for user freedom, control over the product and privacy. On its website, the FSF describes its criteria for its RYF hardware program. "As citizens and their customers, we need to promote our desires for a new class of hardware—hardware that anyone can support because it respects your freedom." The first recipient of an RYF certificate was the LulzBot 3D printer sold by Aleph Objects, Inc.

Recently I read that some surveys indicate that music piracy rates are finally starting to diminish. People aren't ceasing to copy music files because they fear legal reprisal. Instead, music fans are willingly paying for music because finally music companies have gotten better at selling digital music. New albums are quickly released in digital form (not purposefully delayed). And online storefronts are more user-friendly and the download and payment technologies have improved.

## Exclusivity vs. the freedom to innovate

Why do companies spend so much money in legal fees to prevent rivals and customers from making copies of their products? A widely held assumption is that having the exclusive right to make or sell a particular product or brand is key to making a profit. This notion runs deep. Ardent defenders of intellectual property rights insist that stringent enforcement and control makes for the foundation of a strong and innovative economy.

The reality is that it's not that simple. Every industry and product benefits to a different degree. In other words, intellectual property rights are not always necessary for profitability. In fact, some people argue that patents and copyrights actually choke the free flow of ideas necessary to help innovative businesses thrive and prosper.

## RepRap

To learn more about 3D printing and get an alternative perspective on intellectual property, we journeyed to the tall rolling hills of southwestern England to speak with one of 3D printing's most influential people, Adrian Bowyer, creator of the iconic RepRap printer. "Patents do inhibit development—it's unquestionably the case," said Adrian. "It's in the nature of patents that they give a monopoly to whoever holds them for 20 years."

He continued, "James Watt patented various vital aspects of the steam engine. Yet, you look at steam engine development and nothing happened for 20 years during the life of Watt's patent. When that patent lapsed, there was a great flaring of steam engine innovation and then the industrial revolution."

When Adrian and his students created RepRap in 2004, they didn't know it at the time but RepRap printers would become a game-changing technology and intellectual property experiment. The RepRap project emerged from the University of Bath. The ancient city of Bath is near the former home of novelist Jane Austen. Ancient stone arches built during the Roman era circle the city, and restored manor houses punctuate the lush rolling hills.

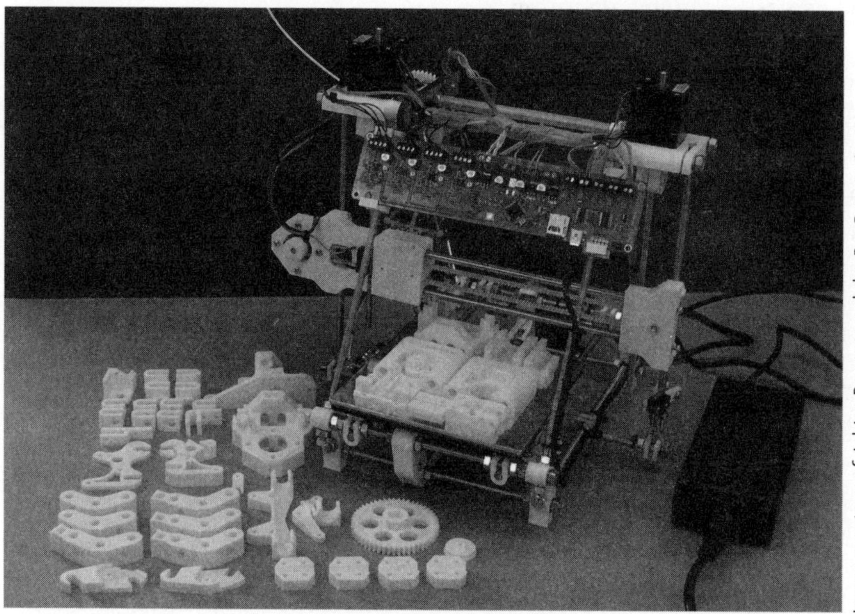

Image courtesy of Adrian Bowyer and the RepRap project

**The RepRap open-source 3D printer can make most of the nonstandard components it is made of, catalyzing its own reproduction and challenging intellectual property paradigms**

Amidst these ancient marvels, the RepRap printer has emerged as a disruptive modern form of product development. Right from the start, RepRap had two characteristics which made it unique. First, Adrian decided to freely share the printer's blueprints on the web using an open-source license. He intentionally did not seek patents on RepRap's design and printing technology. Second, RepRap was designed with an unusual business strategy in mind: its users should never, ever have to buy another RepRap again.

The long-term vision was that each RepRap would be able to fabricate its own replacement parts. This way, spare machine parts could be printed, not purchased. If a user wanted to create a new RepRap (or a thousand new RepRaps), she would simply 3D print out its parts on her RepRap and assemble them herself. Like breeding rabbit populations, replicating RepRaps could reproduce at an exponential rate.

The concept of machine self-replication is reminiscent of the famous Escher print that depicts a man drawing a man drawing a man and so on into infinity, until the figures get too small to see. Most machines, even ones whose blueprints are open-sourced, can't fabricate their own parts. A RepRap printer (full name Replicating Rapid Prototyper) is the physical version of an iterative and endless computer algorithm.

Several forms of disruption could result. If a machine could make its own replacement parts, it would be difficult for any company to claim and enforce patent rights. Another widely used business strategy—planned obsolescence—would also fall by the wayside. Some industries benefit if their products break quickly, but only at the right moment in time. In these cases, engineers calculate how to best design and make a product so it survives long enough to pass out of warranty, but not much longer. It's not your imagination that products break shortly after their warranty expires.

When we asked Adrian his views on patents, he said, "One of the things that many people engaged in so-called conventional business don't seem to understand is that it's possible to have a company that sells something that is completely open and you can still make a profit. People say, 'You don't own any IP. How can you do that?'" He answered his own question: "You have to find a way to add value."

The RepRap project is not against commerce. When he discusses the impact of freely shared machine designs, Adrian made it clear that patents and copyrights should be simply one of several options. In fact, at the time of this writing, Adrian was wrapping up his final semester at the University of Bath and making plans to launch a startup selling RepRap services.

He explained, "The idea that because you give the intellectual property away you can't run a company that makes a profit is demonstrably nonsense. All you have to do is to add value." Adrian's perspective on intellectual property is straightforward: If intellectual property laws aren't benefitting people, then the laws need to be changed.

RepRap and Fab@Home users are free to patent machine enhancements they invent, for example, a special print head that's dishwasher safe. Adrian pointed out that suing one's customers is a crude and ineffective method of protecting one's business model. "Any development or improvement of RepRap design, software, or electronics arises out of its users' own initiatives. There is no central institution giving directions: users themselves invest time and thought in the evolutionary process of RepRap design. If they inspire other users they can all team up and combine their efforts."

The RepRap printer is the embodiment of user-controlled design and production in all senses of the words. Like the Fab@Home 3D printer, RepRap's design blueprints are open-sourced. Like open-sourced software code, instructions for building a RepRap or Fab@Home are freely available online; users don't pay fees or licensing royalties.

Elements from both RepRap and Fab@Home's designs have found their way into commercial models. A notable descendent is MakerBot's Replicator I, one of the most commercially popular consumer-level 3D printers on the market. Today RepRap and Fab@Home printers are found all over the world, some built on the original blueprints and others based on modified designs.

Image courtesy of the Fab@Home team. Photo: Floris van Breugel

**The Fab@Home open-source 3D printer project (2006)**

If these open-source printers have influenced commercial ones, it's worth mentioning here that commercial printing technology has influenced RepRap and Fab@Home. However, the commercial 3D printing companies that held patents that RepRap's design infringed upon never tried to stop the RepRap project (at least as far as we know). Perhaps because the RepRap project did not involve the sale or manufacturing of 3D printers directly—just the open sharing of machine designs. Maybe established 3D printing companies realized that the positive publicity generated by the RepRap project would raise the visibility of the entire 3D printing industry. Or companies figured that if students used RepRap printers in school, they would graduate, get jobs, and eventually embrace high-end commercial printers at work.

What will RepRap's approach, that is, eschewing intellectual property rights and control over machine production and design, do to traditional notions of commerce? Adrian believes that as material goods become much easier to make and more widely obtainable, "The new coin of the realm will be exclusivity." The easier it becomes to make copies of physical things, the more people will strive to own things that nobody else has.

If exclusivity becomes the new competitive edge, the result will be an originality arms race between consumers and between companies. On the one hand, design tools and 3D printing make it easier to create unique and custom goods. Yet, on the other hand, ironically, these same tools make it easy for a unique product to be copied and slightly altered.

In our conversation with Adrian, he pointed out, "Once a new idea comes out, or a version of an old idea that used to be very expensive but now by technical twist is made very cheap, the new idea floods the world thanks to the communications technology we have now." In other words, when everyone can make nearly anything, intellectual property law becomes a crude tool to control the spread of ideas.

In a way, as reverse engineering and copying objects become easier, intellectual property laws become even more crucial to a business trying to maintain a market niche. However, intellectual property rights enforcement—whether by laws or DRM technology—isn't always effective. Adrian summed it up, "Patent law has to labor mightily to have the tiniest influence on economics. When politicians use intellectual law to try to steer an economy, it's like trying to steer a machine with levers made of jelly."

## Open source hardware

The RepRap and Fab@Home projects are good examples of hardware technologies whose machine designs are freely shared. A small but growing number of inventors are open sourcing their hardware designs, releasing them freely under defined conditions of use. Similar in philosophy to open source software, the Open Hardware movement is a response to the limitations of current intellectual property laws.

At the time of this writing, Open Hardware licenses are in their infancy. As the movement develops, the number and variety of available licenses will likely proliferate. Similar to open-source software licenses, an open hardware license gives other inventors permission to use and modify another inventor's designs. Some licenses permit commercial use of a modified design.

One goal of Open Hardware is to create a vibrant, pro-business ecosystem, where machine designs can be freely and safely shared but in a somewhat structured way. Many proponents of Open Hardware licensing models own or work in a business. For many, their profits lie in the sale of enhanced versions of open sourced designs or in service offerings.

The world of software has successfully balanced commercial software and open source software for years. Open source software licenses offer hackers, developers and companies a widely recognized, well-defined alternative to patents, copyrights, and traditional commercial licenses. In fact, many software tools are covered under a smorgasbord of open source software licenses.

In contrast, hardware designers have limited options. Patents are broadly accepted. Yet not all inventors and small businesses who design circuit boards or machines want the cost and legal complications that come with the patent process. Nor do they particularly desire the legal right to prevent others from using their invention.

If inventors choose to open source their electronic hardware design, they make publicly available all the schematics (or a detailed description of needed parts and software, drawings and "board" files)—basically all the information anybody would need to identically re-create the product or object. The published information is made available under a license that dictates acceptable use.

Open hardware can have a guerilla ethos to it, if inventors so desire. Once a design blueprint is open sourced, it becomes what's known as "prior art." Prior art, or previously published technical information about an invention, makes it difficult for non-inventors to claim ownership to the idea and get a patent for it.

The set of licenses being defined by the Open Hardware movement is in its infancy, nowhere near the maturity of open source software. However, the use of Open Hardware is increasing, thanks to a passionate community and the vision and tenacity of two young women, Alicia Gibb and Ayah Bdeir. In a community that's largely populated by men, Alicia and Ayah stand out as leaders, technologists and entrepreneurs.

In an interview with *Make* magazine, Alicia described Open Hardware as a catalyst to technological innovation.

> *I think open source hardware is the 21st century patent system . . . Open source hardware also is about open sourcing innovation, but also democratizing innovation, but does not come with 20 years of exclusive rights. The benefit is that you have an entire community contributing to your designs, innovating, and sharing their derivatives to your product. It pushes the original designer to create a better product and continue to improve it rather than lock it in a 20-year stalemate.* [9]

Open-source hardware might not be the solution for every kind of invention. Open-source philosophy works well for relatively complex products, like software and electronics, where the original creator's expertise is still useful after the source has been opened. Relatively few people can improve someone else's complex circuit or algorithm, so the original designer usually still stays in control of the project even after it is released.

A simpler invention, however, is too easily copied and the creator forgotten. Its inventor likely won't be able to benefit from adding value if there's no need for expert services—this is especially true for simple, elegant inventions that are "obvious in hindsight." I wonder whether this is why some of the most vocal advocates of open source licenses are software developers and people who invent complex electronic devices. Whereas people who invent a new recipe for a material or a new mechanical device often continue to rely on patents.

## Micropatents

As 3D printing technology develops, Open Hardware licenses may be a good alternative to intellectual property law and digital rights management. Another possibility could be to adapt our current system of intellectual property law to make it friendlier to small businesses and solo inventors. Consider that in low-income countries, micro loans give low-income individuals the opportunity to found a small business. In the development of new products, micropatents could level the patent playing field between large and small businesses.

Similar to a regular utility or design patent, a micro patent would be formally issued by a government patent office. However, a micropatent would be a smaller "unit" of intellectual property than a full patent, hence cheaper and faster to get. Maybe the patent office could impose a mandatory limit on patent scope to qualify as a micropatent.

Here's how micropatents would work. For a few hundred dollars an inventor would submit a document describing their invention to a centralized government micropatent repository. The submitted document would be time-stamped and immediately publically released. A micropatent's lifespan would be shorter—maybe 5, not 20, years. Like a trademark, a micropatent would protect only objects sold commercially to prevent trolling. Like a patent, a micropatent would cover a utilitarian application. Like a copyright, it would be easy and straightforward to obtain.

Some argue that a glut of micropatents would lead to patent gridlock and fragmentation of the intellectual property space. Their 5-year timespan would make this unlikely. In fact, a micropatent's narrow field of use would make it easier for other inventors to work on parallel inventions. The requirement that a micropatent must be used commercially to remain valid would make it difficult for companies to use a war chest of micropatents to block their competitors.

In fact, micropatents would improve the ratio between legal processes and practical application. After all, much effort is put into evaluating and obtaining patents, yet most issued utility patents are rarely commercially used. In addition, only an estimated 0.1 percent of patents are ever litigated to trial, and in those cases, half are invalidated.[10]

In contrast, if patents were narrow in scope and quick and cheap to obtain, such an approach would de-emphasize the patent prosecution process and instead, would emphasize the practical commercial value of intellectual property protection. Once an inventor was granted a micropatent, the inventor would immediately be granted an implicit, short-term exclusive right to the disclosed idea. Only in case of a dispute by two *practicing* commercial entities, would allegations of patent infringement be evaluated by traditional tests of novelty, utility and non-obviousness.

## Striking the right balance

Where is all of this headed? If the truth is stranger than fiction, maybe a good place to see what lies ahead is in a science-fiction novel.

In his novel *Makers*, science-fiction author Cory Doctorow paints a riveting picture of a future world where 3D printers have become commonplace. The world depicted in *Makers* is one that's suffering from hard times. Unemployment is high. Large segments of the population have fallen off the proverbial grid and are living in self-organized shantytowns.

The action begins as two of the book's main characters, Perry and Lester, inspire a small-scale economic revolution when they teach local people to design, 3D print and sell their own wares. At first, optimism reigns. As economically marginalized people learn to design and 3D print products, work collectives spring up all over the country and world. People gain employment and get back their self-respect.

As the story continues, entrepreneurs make a living selling some truly disturbing 3D printed custom artifacts. In a daily flea market that takes place in an enormous parking lot of a bankrupted Wal-Mart, entrepreneurs casually hawk printed bongs, single-shot zip guns, and patterned contact lenses. Teenage vendors sell 3D printed biological oddities known as "bio," for example, a bracelet made of embryonic stem cells grown into little balls of fur, bone, and skin. In this future world, another typical bioprinted goodie is artificial roses printed out of delicate leather made of cultivated fetal skin.

In the second half of the book, economic dystopia sets in. The 3D printed economy comes crashing down as businesses fall apart under an onslaught of high-speed counterfeiting. New products have a diminishingly brief window of profitability before low-cost 3D printed replicas undercut the market. The Disney corporation, in search of new profitable ventures, kicks off a series of aggressive intellectual property lawsuits that eventually cripple what's left of the economy.

*Makers* raises fascinating legal issues about a future world that may not be so far off. From one perspective, the upside of democratizing the power of production is that disenfranched people gain personal empowerment. The downside of widely available tools of production is an economic meltdown. Rampant counterfeiting erodes profit margins. Add to that the devastating effect of unrealistic intellectual property laws that attempt to enforce private ownership of ideas in an era when anybody can 3D print exact copies of anything.

The solution is to create realistic and balanced legal frameworks. As eloquently expressed in an article about molecular manufacturing by the Center for Responsible Nanotechnology, "Some of these risks arise from too little regulation, and others from too much regulation. Several different kinds of regulation will be necessary in several different fields. An extreme or knee-jerk response to any of these risks will simply create fertile ground for other risks."[11]

The legal challenges introduced by 3D printing flicker on the edges of public consciousness. We're just now beginning to explore this new frontier. It won't be long now until the law saunters in to see what's going on.

# 13 Designing the future

A popular way to describe the potential of 3D printing as the ultimate manufacturing machine is to compare it to the Replicator in *Star Trek*. The Replicator was a machine that took verbal commands and fabricated whatever Enterprise crewmembers requested. In other words, the Replicator was a machine that had the power to make anything.

## Tea. Earl Grey. Hot.

When I watched *Star Trek* as a kid, it used to frustrate me that nobody ever put the Replicator to the test. On episode after episode, whenever the Replicator appeared, all anyone ever asked for was . . . a cup of Earl Grey tea. On a daring day, a crew member might ask for a piece of cheesecake. Perhaps Mr. Spock's lack of imagination could be excused because of his Vulcan heritage. But not that of his human colleagues.

As an adult, I place the blame on *Star Trek*'s scriptwriters. Given a machine that could make anything, their imaginations could only stretch as far as having this wondrous Replicator spit out a mundane cup of tea. In my lab, we call this phenomenon the "Earl Grey Syndrome."

Design for 3D printing suffers from the Earl Grey Syndrome. Like the unused power of the Replicator, 3D printing offers us an unexplored new design space. Yet our imaginations remain enslaved by past experience. We humans are creatures of habit. Our creations are elaborations of what we're already familiar with. Like the old saying goes, "If all you have is a hammer, then everything looks like a nail."

Part of the problem is the design software we have to work with. Computer-aided design tools are critical in the 3D printing process, yet design software remains entangled in limitations imposed by once-inadequate computing power. In addition, today's design tools were built to work within the physical constraints imposed by conventional manufacturing machines.

To break out of the Earl Grey syndrome and take full advantage of 3D printing technologies, we humans need to step up and do better than *Star Trek*'s scriptwriters. Back on planet Earth, a bit closer to home, I see the Earl Grey syndrome play out in class. Its roots run deep.

I teach an undergraduate product design course. In the beginning of one class, I asked students to design a pencil holder that will be both practical and visually interesting. I explained that their pencil holder project designs would be 3D printed, so I spent time teaching them about the new design possibilities laid at their feet.

When I assigned the pencil holder project, I urged students to cut loose with their designs, to make their pencil holder as "wild looking" as possible. "Think outside the box—literally," I told them, hoping to see some mind-bending, avant-garde designs.

After I gave this assignment, my students returned to class with their designs complete. As I walked around the room and looked at their computer screens, my hopes were dashed. Their pencil holder designs were mostly well-engineered. They could remain upright. Most holders could hold pencils and pens. In fact, most students turned in good, solid designs that would end up being sensible and good-enough pencil holders.

A few bolder souls designed some surface flourishes to liven up their pencil holders. Most of the submitted designs, however, were timid variations on what already exists. Undergraduate engineering students aren't the only ones who suffer from the Earl Grey Syndrome. Outside the classroom, I've noticed that even professional designers frequently forgo the freedom of creation afforded by a 3D printer.

Marshall McLuhan, Canadian philosopher and author, aptly described the situation, famously proclaiming "We shape our tools and thereafter our tools shape us." McLuhan's powerful insight helps explain the design myopia that characterizes the Earl Grey Syndrome. A few decades ago, we shaped computer-aided design tools that respected manufacturing constraints that no longer exist. Yet, today, these design tools continue to shape us. To break this stalemate, we must re-shape our design tools.

There are two opportunities, in particular, that I find promising. First, making design software intuitive and fun to use. Second, improving the way computers "think" about shapes, because the way computers think about shape dictates, to a large extent, how far they can let us explore.

## A bicycle for our imagination

In a 1990 interview on public television, Steve Jobs described computers as "the most remarkable tool that we've ever come up with. It's the equivalent of a bicycle for our minds." His point was that if given technology to boost basic human capacities, humans can dramatically extend the limits of what they are capable of. Unfortunately, for most of us, design software has not yet become a bicycle for our imaginations.

Imagine this future scenario: a nine-year old girl gestures in the air. On a nearby screen, a design slowly takes shape. On the girl's fingers are small stickers that have sensors that wirelessly convey information to the design software. The software receives the sensor data and interprets the girl's motions as she smooths, selects and pinches her onscreen object into the right form without ever touching a keyboard or mouse. The computer understands her intentions and, like a master artisan, grows the design into its perfect final form.

If only design software were this easy to use. To fully unlock the power of 3D printing technologies for everybody, design tools must become more intuitive to learn, more fun to use, and more capable. One promising approach is to make design software that looks and acts like a video game.

### Gamifying CAD

Several months ago I received an e-mail from Eric Haines, a veteran software developer who studied computer graphics as a master's student and has had a long career at Autodesk. Eric thought I might be interested in a new online world called *Minecraft*. "You could describe *Minecraft* as an 8-bit vision of the future of design for 3D printing," Eric told me. "Millions of people play *Minecraft*. It's adult LEGOs."

The reason Eric had contacted me was because he thought I might be interested in *Mineways*, a software tool he created.

*Mineways* is an open source (and free of charge) software tool that enables people to edit, then 3D print what they've designed in *Minecraft*. Eric developed *Mineways* when he realized the rich virtual worlds people were building online in *Minecraft* were begging to be materialized in physical form. Eric wrote *Mineways* in 45 days in his spare time and gifted the software to the world on Christmas Eve, 2011. Since then, *Mineways* has been downloaded more than 50,000 times.

Eager to learn more, I arranged to meet Eric so he could show me the online world of *Minecraft*. One of the first things I noticed in Eric's living room were 3D printed tchotchkes scattered around. He showed me a 3D printed replica of a castle, plus a blocky, rough-hewn, gray stone house with a yellow thatch roof. The house, Eric explained, was a printed replica of a village in *Minecraft*.

We sat down in front of a widescreen computer monitor, and Eric logged into his *Minecraft* account. Eric explained that *Minecraft* is a multiplayer game where players create an avatar and set up their own custom-designed virtual world. By clicking and dropping cube-shaped chunks of raw material into place, players create elaborate and rich fantasy worlds. At the time of this writing, *Minecraft* had an estimated 8 million active players, a population the size of a small European country.

I peered over Eric's shoulder at the screen as he walked me around his virtual world, Vokselia. "When you start, you get an account and then you get a world with nothing in it." His *Minecraft* world, shared among a few friends, boasts houses, castles, a giant glass dome, even a few farm animals. Eric showed me one of his most ambitious designs, an angular grasshopper. He explained that it was a challenge to design the grasshopper's spindly legs and antennae in such a way that they could survive the 3D printing process.

In a *Minecraft* world, each cube-shaped building block represents the real-world equivalent volume of one cubic meter. Unlike contemporary video games that are so sleek and finely rendered they look like movies, the graphics on *Minecraft*, at first glance, appear crude, even primitive. *Minecraft* worlds are made of big, blocky chunks of raw material. Trees, houses, lakes, structures are coarsely pixelated. Each flat surface appears to be made of individual tiles.

What makes *Minecraft* so addictive? Part of the appeal is that the visual effect is magical, the cube-shaped building blocks lend a mystical air to the cities, buildings, and fantasy worlds that people build. "In video games, the graphics may be higher quality but you can't change and move things," said Eric. "Most video games are interactive but you can't get to the level of design you can here."

This fantasy world has strict and consistent internal rules. The game imposes physical laws onto players so their design space is a blend of fantasy tempered by the discipline of the physical world. Players must navigate their way through the same constraints that introduce real-world complexity into the building process. Since an elaborate design project can take a lot of time, to speed the game along, a day in *Minecraft* lasts only 20 minutes.

Eric showed me a train station's floor tiling, a mosaic of browns, reds, and yellows. To make brown tile, *Minecraft* players can't just choose "brown" and apply the color to a floor tile. Instead, there's an organic and laborious process to make brown tiles. "These tiles are colored from special dyes for wool. You have to go shear sheep and gather wool to get these colors," said Eric. "Pink is quite rare. Brown is quite rare. You have to find cocoa to make this brown color tile."

I suspect part of *Minecraft's* appeal is that players can work together in teams on massive building projects, creating the sense of old-fashioned community. Somewhat like a barn-raising, Eric and his friends teamed up to build a virtual masterpiece, Community Station. This train station was modeled on St. Pancras Station in London. Each player contributed time, raw materials, or design expertise to the project. When I saw Community Station, there was a sparkling Christmas tree in the corner of the station's atrium—apparently the players hadn't yet gotten around to taking it down after the holidays ended.

*Minecraft* demonstrates a new design paradigm: Gamified CAD. For $27, *Minecraft* offers players easy, powerful design tools, plus an online community of fellow players and builders. Back when I was a young engineer, learning to use design software was almost like learning to manage an airport control tower. It had a bewilderingly complicated user interface, special vocabulary, and was way beyond the skill of the average user. It wasn't fun to learn at any age.

*Minecraft* is so easy an 8-year old can play. In fact, a few months ago, my 8-year-old son started playing *Minecraft* at home. After a few days of learning his way around, he designed and built me my own virtual home (near his tree house), complete with a shower stall and bed.

*Minecraft's* physical constraints seem to fascinate him. He set up several experiments to test the explosive power of TNT, blowing massive angular craters into dirt and cliff face. My son explained to me that a few chickens died when they accidentally wandered near the explosion; yet somehow a passing cow survived the blast and shortly afterwards, was grazing peacefully near the gigantic crater.

Deep inside *Minecraft's* software (of little interest to anyone other than a few computer scientists) lies an interesting design architecture. Its software does not use industry standard design models. *Minecraft's* blocky graphics are not generated by traditional solid modeling (the design tool of choice for engineers), nor surface mesh (the design tool of choice for animators and artists). The heart of *Minecraft* is a digital unit called a voxel. One digital voxel translates smoothly into a precise physical location in the volume of a three-dimensional object.

Voxel-based design chews up a lot of memory but it appeals to users conceptually as a design unit, or digital brick. According to Eric, voxels, though seemingly crude in appearance, offer regular people an appealing and straightforward way to design. "The voxel as a unit is something that people can relate to very directly. For years—decades—I've been working on surface modelers, where you draw polygons or curved surfaces. But it's sort of a nebulous, counterintuitive. If you don't have training on how to use surface mesh, it's hard to use, it's not how people think."

Eric showed me a beautiful *Minecraft* creation, a majestic aerial depiction of a Spanish city created by Lee Griggs and Tomás Fernandez Serrano at a graphics design company called Solid Angle. In their *Minecraft* world, Lee and Tomas designed a domed building that presides over a circular intersection of six streets.

Image courtesy of Lee Griggs (Solid Angle SL) and Tomás Fernandez Serrano (forominecraft.com). Rendered with the Arnold global illumination renderer.

**A *Minecraft* scene created by designers Lee Griggs and Tomás Fernandez Serrano. If you look closely, you can see the Mario Brothers in red and green hats to the left of the central dome.**

The model city surrounding the domed building is sprinkled with small red-tiled roofs. In the distance you can see a colosseum.

As you can see in the picture above, the original image of this scene depicted the figures of the Mario Brothers climbing atop one of the buildings. However, in a hint of the intellectual property challenges that loom ahead, the copyrighted figures of the Mario Brothers were removed when this image was selected to appear on the cover of a design magazine. The magazine—concerned about a potential copyright challenge from Nintendo—asked that Griggs and Serrano take the Mario Brothers out of the picture, just to be safe.

When my classmates and I were learning to use CAD in the late 1980s, we would have scoffed if our teacher had told us that our children would someday happily play with simple computer-aided design tools. We didn't have the computer power in those days to run a richly detailed virtual world. Nor did we have access to a 3D printer to fabricate our designs.

## Matter compilers

What if you had a modified version of a Replicator that didn't just obey simple verbal commands? This machine would be smarter. I'll call this intelligent machine "the Extra-smart Design Assistant." You would tell your Extra-smart Design Assistant the problem you were trying to solve, not just the name of an exisiting object you were trying to replicate.

Here's an example. Instead of commanding your Extra-smart Design Assistant to "make a new support bracket for my bookshelf," you would describe the problem: "I need a new bracket to support the weight of a set of books." Next, you would tell the computer what this bracket would need to be capable of, or give it design requirements that some people call "design specs." Like a good designer, your automated Extra-smart Design Assistant would listen carefully to your requirements and come up with a solution optimized for its tasks.

To design your shelf brackets, you would verbally rattle off several design requirements. "The bracket is intended for a bookshelf that's 6 feet wide, will hold heavy textbooks, and will attach to a vertical wall. In addition, I want the designed bracket to weigh as little as possible but be capable of bearing a load of at least 50 pounds of weight." Finally, you would tell your Design Assistant that you plan to 3D print the bracket out of a hard plastic. Your Design Assistant would carefully think, then after a bit of calculation, would display its suggested design solution.

In my lab, we created a software tool based on this concept. The tool ran algorithms to calculate the optimal design for a 3D printed bracket. We punched in the design requirements and pressed "Enter." We were startled by the results.

The computer came up with a novel design for a bookshelf bracket that looked nothing like the right-angled brackets you can buy in your local home supplies store. Its suggested design was a fibrous tangle of materials wrapped around a few beautifully shaped internal cavities. I don't think that I could have designed such an organically-shaped (and optimized) design using today's software design tools. There's no way I could have made this bracket by hand, given the weight and material capacities the computer had to consider.

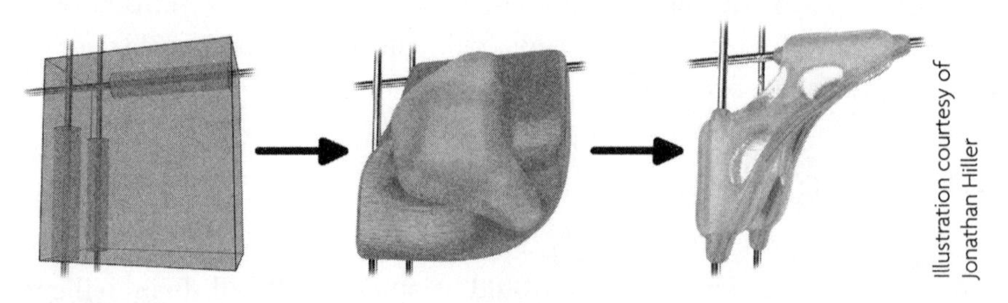

Illustration courtesy of Jonathan Hiller

**Someday design will be done by matter compilers that generate the best structure for a given objective and then 3D print it. Starting with volume constraint (left), the compiler produced an optimal bracket to hold three bars.**

What will happen in the future, when intelligent computers learn to bridge the gap between what humans need and what multi-material printers can make? The result will be next-generation design software, or what I like to call a "matter compiler."

The term matter compiler was coined by science fiction author Neal Stephenson. In his novel *The Diamond Age*, the book's characters tell their matter compilers what to fabricate and in a moment or two, they'd pull a plastic mattress or food or a firearm from the machine. Stephenson's matter compilers were not 3D printers, of course; they were "nano-assemblers" that re-arranged atoms streamed from a central "feed." Nor could they design new things; like the *Star Trek* Replicator, they were limited to replicating what already existed.

The true promise of a matter compiler is that it would combine the power of artificial intelligence with 3D printing. First, the artificial intelligence would apply algorithms to automatically "compile" high-level human-provided requirement into an optimal design, suitable for 3D printing. The 3D printer would then realize the design physically. Put together, such a matter compiler would enable us to design new objects that do not yet exist.

Perhaps a more adequate analogy to our kind of matter compiler is a traditional software compiler. A software compiler translates a high level computer language like C++ into machine code optimized for performance on specific computer hardware. The machine code is then executed by a physical processor. Similarly, our matter compiler would translate high level requirement into a design file that takes advantage of the latest materials and features of your 3D printer. The printer would then build the object described by the design file that the matter compiler generated.

Perhaps a wall bracket isn't the best example since it's a simple design challenge. The power of a matter compiler attached to a 3D printer would become apparent if you were trying to design and make a complicated machine. If 20 years from now I was invited to design a robot for the next Mars mission, here's how I'd like to do it.

First, I would describe the unique conditions on Mars. Then I would give some overall dimensions, for example, a weight range, what raw materials will be used, and so on. Finally, I would provide the performance specifications: its speed, its stability, its efficiency, and so on. Then I'd hit "Enter" and wait for the matter compiler to generate some new design ideas. Someday, when matter compilers become an everyday design tool, the time-consuming process of trial and error that characterizes complex design projects will be handled very quickly.

## Interactive evolution—breeding designs

As intelligent as it would be, even a matter compiler wouldn't be able to read minds. True, the compiler could devise great design solutions based on what you tell it you need. However, it could not brainstorm with you. What if you and your design tools could converse and together work through several different iterations of a design-in-progress?

If a computer could quickly toss out design ideas in rapid iterations, you could pick the ones you like. Then, the computer could "study" the design suggestions you chose and make some rapid adjustments to them and offer them back to you. You could again select your favorite options, return them to the computer, and again, it would make adjustments and offer you the result yet one more time.

An iterative design process in which a human and a computer bounce ideas back and forth is called interactive "evolutionary design." Like biological evolution but much faster, a computer can reconfigure a design using mathematical algorithms. The beauty of interactive evolution is that the user does not need to know anything about the inner workings of the computer's process. A user does not need to know how to use design software or how to calculate optimal design solutions. In fact, somewhat like an artist browsing the web for inspiration before beginning a new project, interactive design software could inspire new ideas by provoking the designer to venture into design solutions that no one else has yet thought of.

We made a working prototype of interactive design software in my lab. A former student, Jeff Clune, created the software. Jeff and his team made the software available on a website which he named Endless Forms, a play on Darwin's famous description of the process of biological evolution.

Here's how Endless Forms works. People breed dogs by crossing dogs that have certain characteristics they want to appear in their puppies. Endless Forms allows a user to "breed" designs by choosing the ones they like. In other words, the Endless Forms' website brokers a rapid back-and-forth exchange between the user and the interactive design software.

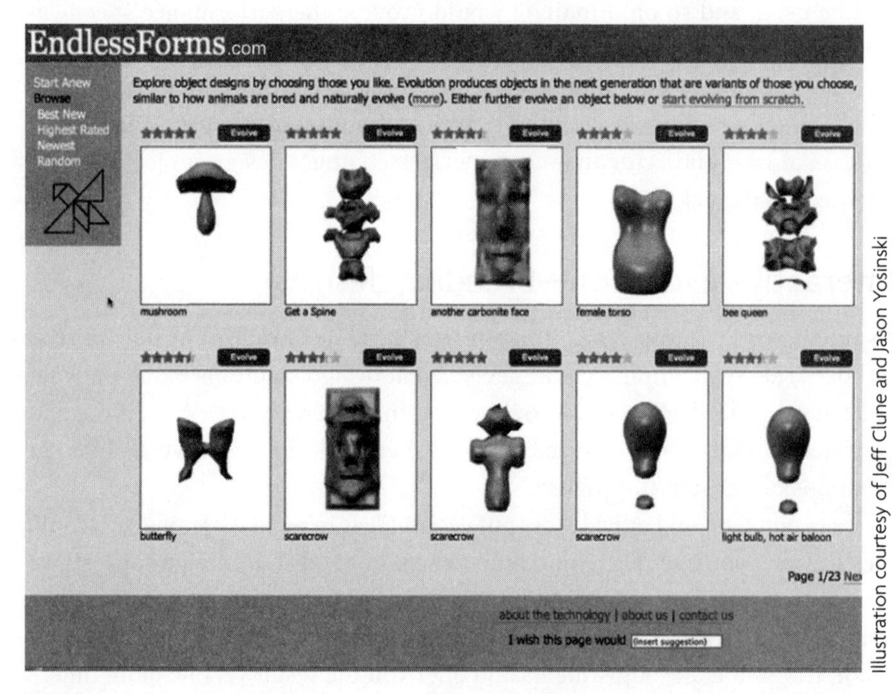


Illustration courtesy of Jeff Clune and Jason Yosinski

**The Endless Forms website allows users to choose designs they like and refine those designs over several iterations.**

Imagine that you were assigned a design project, a new perfume bottle. You don't know how to use design software and you would prefer not to base your new design on an optical scan of an existing bottle. Instead, you would sit down at the Endless Forms website and make your first design choice by clicking on a crude shape from a collection of several basic variations on the website. In response, Endless Forms would note your first design choice, "think" about it, and rapidly make a few calculations. It would then generate several new shapes that were somehow similar to the first shape you chose.

Next, you would choose again from this new set of shapes. Endless Forms would repeat the process, noting your design choices and generating a new batch of related shapes for your perusal. Eventually, after several rapid iterations and rounds of shape selection, you would start to see a design emerge. When you felt that finally, the design offered to you was what you wanted, you could save the final design and your design project would be completed.

When Jeff and his students set up the Endless Forms website, they realized they could watch peoples' design process, first hand. When users are given a nearly blank slate and allowed to make a series of rapid selections, interesting designs emerge. In just a few months after the site went live, users generated over three million oddly shaped objects from lampshades shaped like mushrooms, fertility goddess figurines, to cubes featuring spooky faces jutting out of them.

When people use conventional design tools, the interface and software features seem to shape their thinking, maybe by making people self-conscious. In contrast, on Endless Forms, users "bred" designs by impulsively clicking on a shape they liked and doing that again and again. The process of rapidly choosing and then sitting back to wait, combined with the fact that the computer did most of the work for them, seemed to bring out a new expression of people's subconscious.

Endless Forms may not be the best design tool to create machine parts for the next Mars Rover. When users create a design by impulsively pointing at appealing shapes, the result will likely not be an object that's honed and mathematically optimized for maximal performance. Instead, as an early prototype of a new type of human/computer interaction, perhaps the value of the Endless Forms project was that we proved that it is possible for a computer and a human to design "collaboratively."

It turns out that it is far easier for humans to critique design options that are offered to them than it is for them to design from scratch. While you probably can't design a house from the ground up, you can easily point to houses that you like and don't like. A good architect will listen to you and from your feedback, and within a few iterations, produce a house to your liking. People bred dogs for centuries without understanding animal genetics; farmers bred corn long before they understood plant genetics. We can breed complex things without understanding how they work.

Endless Forms demonstrates how computers and interactive design software introduce a new dimension to creative work. Two pioneers of computer art, Stephen Todd and William Latham, describe a computer-based artistic process as one that takes place in "two stages: creation and gardening. The artist first creates the systems of the virtual world, applying whatever physical and biological rules he chooses: light, color, gravity, growth and evolution, and other rules of his own invention. The artist then becomes a gardener within this world he has created; he selects and breeds sculptural forms as a plant breeder produces flowers."[1]

## The language of shapes

Getting human designers and computers to creatively and seamlessly work together remains one of the biggest challenges in design for 3D printing. A related issue is less visible to our human eyes, but still as essential. To make better use of a 3D printer's production capacity, computers need to learn to improve how they "think shape."

Humans parse, record and often think about reality by capturing it in spoken and written language. A culture's language reflects its values and physical environment. Skilled speakers use complicated grammars and deploy a massive and colorful vocabulary. Other speakers loosely string simple words together in a crude and poorly organized, ungrammatical skein of words.

Instead of using language, design software parses and records external reality by capturing and keeping track of data that describes a physical object. A more technical term for the way design software digitally captures an object is as a "geometric representation."

Design software of the future, like a child learning language, will gradually become more fluent in depicting shapes. The successful development of a child relies on it learning to speak smoothly in a way that's responsive to the situation at hand. Similarly, the successful evolution of design tools will depend on whether we can create software that can rapidly and accurately evolve new geometric representations in response to the design problem at hand.

Some human languages have words to describe emotional states or particular situations that other languages don't. The complexity and available vocabulary of a human language influence what can be said. Similarly, the way a particular design software tool handles internal geometric representation determines how well it can manipulate a design.

Most design software today speaks a simple form of language and thinks like a traditional paper blueprint. It speaks a simple crude dialect of simple words. For example, design software has two words to describe the presence or absence of material in a design: "there" or "not there." There's no capacity for ambiguity. There's no way to describe a physical object's growth, or change or conditional adaptation. Like a non-native speaker or a child learning his first words, design software simply and baldy states the present details of a design's shape and materials.

Until design tools can speak more eloquently we can't fully tap into the potential of 3D printing. There are simple ways for a computer to "think shape" and there are more sophisticated ways. If I had to rank design paradigms from simplest to most fluid and adaptive, here's how I would rank them.

The simplest geometric representations would be traditional paper blueprints, surface mesh models or solid geometries. These design notations describe a simple, fixed shape. They are the equivalent of speaking a few simple descriptive words.

Next, somewhat more adaptive is design software that can handle parametric designs. These allow a user to define generalizable geometry that is adjustable according to a few parameters. For example, a family of hammer-like shapes could all be described using a single geometry with various lengths and widths of their handles.

At this point, we move into the future realm. The following types of design languages are highly experimental, found mostly in research labs and other cutting-edge design experiments. In one approach called "design as programming," a computer describes a shape as a sequence of steps in a particular order, somewhat like describing a cake by its recipe, not by its final appearance.

Next, a more complex approach is offered by what we call "generative systems." Such systems literally "grow" a shape from a seed, according to a given set of rules.

Finally, the most sophisticated and fluent future design representations will be "reactive blueprints." These are designs that modify themselves to fit the conditions where they will be used. Like an eloquent orator improvising on a podium in response to the crowd's mood and questions, a reactive computer blueprint software is dynamic.

The latter is how nature works. A plant's DNA does not specify the plant's final shape explicitly. It specifies a set of rules that will govern how the plant grows in response to any particular condition it may find itself in.

Until the advent of 3D printers, complex ideas about representation of form were fodder for fantasy, theoretical mathematics, and computer graphics. Only nature had the manufacturing ability to physically make the complex forms described by a generative blueprint. As 3D printing technology improves, these new design concepts will finally be able to step out of the virtual and into the physical world.

## Design as recipes

A recipe may be an appropriate way of thinking about design software of the future. A recipe for a cake does not describe the details of the shape and composition of the final product. Instead, a recipe describes the process of making it as a sequence of steps, essentially, as a procedure, a program that if executed meticulously will result in the desired outcome.

A relatively simple recipe can result in a rather complex outcome. Sprinkle sliced apples and raisins on puff pastry, roll it up and bake it, and you have an apple strudel. The sequence of steps listed in the recipe are substantially simpler than a narrative verbal description of the strudel's exact geometry, appearance and material composition. Yet, the recipe creates an object much more complex than its simplicity would suggest. In other words, the whole is greater than the sum of its parts.

The equivalent approach to a strudel recipe in design software is a paradigm called "geometric programming," or "function representation." Geometric programming requires a different kind of thinking, a different kind of imagination, and a different kind of designer. The resulting objects are far more complex than what can be designed by a traditional point-and-click design tool.

In geometric programming it's easy to describe repeated structures and semi-periodic structures that vary a bit but are mostly the same. Hierarchical structures composed of smaller substructures can also be described.

For this reason, geometric programming offers a human designer great efficiency when designing patterns that are complex or made up of many small parts. For example, imagine if you had to design a suit of chain mail made of extremely fine mesh by pointing and clicking millions of tiny rings. In conventional CAD software, this would be an excruciatingly laborious and time-intensive. task.

If you were using a geometric program, however, this task would be easy. You would simply write a "recipe" as follows: create one ring of 3 millimeter diameter and ½ a millimeter thickness. Then, duplicate the ring 1,000 times in an entangled row. Third, duplicate the row 1,000 times vertically to create a cloth. Press "Enter," and your CAD program would follow the recipe to produce your fabric.

If the weight or density of the fabric had to be tapered slightly, you could simply adjust the recipe. This time, you would tell your computer that, as it duplicates each row, the next row of mesh should be, say, 1 percent smaller in size than the preceding row. You could specify that after doing this for 2,000 rows, then the size of each tiny mesh link should slowly increase again, perhaps increasing in size by 0.5 percent. You could even change the blend of plastics as you go.

## Generative design

Generative design takes the idea of a recipe one step further. Instead of a centralized script, generative design uses a seed shape and a set of rules that specifies how that shape should develop, or unfold, over time. For example, to grow a complex, tree-like shape, you would begin with a simple "seed": a simple vertical cynlinder attached to a flat base. Next, you would specify one rule: Attach two more 10 percent smaller cylinders in the shape of the letter "Y" at every free end of each cylinder. Applying the rules recursively to the seed would cause the seed cylinder to grow into a large tree-like shape.

Image courtesy of Nervous System, Inc.

**Generative designs can be used to design objects with a more organic look and feel.**

Generative design systems are now just starting to flourish commercially, but much of this work is done by skilled designers for custom projects. Generative design programs are sometimes called "shape grammars" and when printed, result in sophisticated three-dimensional patterns.

Generative systems can be used with different rule sets, different basic operations and basic shapes, different syntax, and so on. Some people experiment with a set of fixed rules. Others use conditional rules that depend on the shape's current state. For example, a rule might specify to split the cylinder into smaller cylinders only if the cylinder is larger than 5mm in diameter. One variant, named after biologist Aristid Lindenmayer, are sets of rules that are famously known for their capability to generate plant-like shapes and organisms, often used to generate lush natural computer-generated graphics scenes in movies and video games. (You didn't really think there was an artist drawing each tree, did you?)

Generative design software can handle semi-random rules that may or may not trigger a shape change depending on chance, or may generate different outcomes according to a pre-defined probability. Such semi-random rules make sure that every tree in a scene looks like a tree, yet also every tree is slightly different. Generative representations could go on and on, and the shapes become richer and more complex, not unlike biology.

Biology's representations are generative: our DNA does not encode where each cell in our body will be and what it will do and how each of the neurons in our brain will connect. Nor is a DNA a script that when executed generates a body. Instead our DNA encodes a set of rules applied iteratively to the first germ cell until a full body is generated.

## Reactive blueprints

The most sophisticated form of language to manage the description of shapes is one that's dynamic and responsive. I call this sort of design process "reactive blueprints." Reactive blueprints enable a designer to automate the production of sophisticated shapes that will adapt "in real time" to their environment whose specifics are yet unknown to the designers.

Project 'DIGITAL VERNACULAR' by Shankara S. Kothapuram, Mei-ling Lin, Ling Han, Jiawei Song, part of design studio: 'Machinic Control' led by Marta Malé-Alemany and taught with Jeroen van Ameijde at the Architectural Association School of Architecture, Design Research Lab (DRL)

**Someday reactive blueprints will design homes whose shape is ideal for their environment.**

Reactive blueprints don't do small jobs. For example, you would not use reactive blueprints to 3D print a replacement knob for your washer at home.

Instead, reactive blueprints would be applied to design projects where the environment is unknown and even changing.

A reactive blueprint could guide a large concrete printer that's printing out a house that needs to adapt to a yet unknown terrain, a bridge that needs to adapt to wind conditions, or a lampshade that needs to compensate for particular ambient lighting conditions. A reactive design blueprint would be ideal in surgical chambers. Maybe someday in situ 3D printers will fabricate biological tissue inside the human body that adapts to the unique conditions of every individual body.

Design software executing a reactive blueprint would first need to scan the target environment. It would also need to be able to *simulate* the target environment with great accuracy in order to know which rules will be triggered and when. Using this information it will be able to "grow" the shape suitable for that specific environment. The shape would then be made by a 3D printer.

When a designer is preparing a 3D printer to print out reactive blueprints, it will be impossible to know in advance exactly how the final design will turn out, and what it will look like. The feedback for each situation will be different, changing the specific details of the ultimate design. In nature, for example, plants often grow toward a light source, but after a period of growth will stop at a certain height due to internal stress induced by their own increasing weight. If you were given the ability to design a plant's growth, you could set up rules that permit the plant additional growth when it is exposed to external light. You would temper that growth with rules based on stress sensors that manage the plant's growth so it doesn't get too tall and cause internal loads that are too high.

Working together, these two rules—external and internal—if applied to plants in different conditions, would result in plants of different heights and shapes. Even two plants with the same rule set, if one were kept in dim light and the other in bright sunlight, would end up looking different because of the different feedback their sensors provided and the different response that feedback triggered from the rules.

## One lampshade design, many custom lampshades

How could future designers apply reactive systems to the product design process? Imagine a dynamical blueprint for a lampshade. The shape of the lampshade would be specified using a set of rules that would be applied to a simple "starter shape" to develop the lampshade's final design. You would place the lampshade in a corner of a virtual "room," next to a virtual "window" in a computer simulation. The software would calculate feedback on the level of light available next to the virtual window would steer the design into a particular shape.

If you were to place the lampshade in a different virtual room next to another light fixture, the lampshade's design would evolve into a new shape. By applying computer simulated environmental cues to the design process, reactive blueprints specify the product not by describing a desired shape, but by letting it form in response to data.

One lampshade design could lead to many other custom designs, each adapted to its particular use, but each a lampshade nonetheless, like individual trees in a forest. If you had this reactive lampshade blueprint, you could generate a different lampshade design for each room in your house, provided that you could accurately simulate the lighting conditions in your house in your design software.

## Printers that think

Taking the idea of dynamical blueprints even further, design feedback could be fed directly into the printer—not from a computer simulation, but from reality. Imagine a 3D printer that would know what it needs to print and adapts to the conditions of the print. Most 3D printers are "blind" in that they execute their instructions and don't look to see if the instructions yielded the target object. Such systems are called *open loop*.

Daniel Cohen at Cornell explored the idea of 3D printers that keep an eye on their output. Daniel created a closed loop printer, one that "watches" the outcome of its printing and adjusts dynamically in response to various situations. For example, a printer prints with some oozing material that doesn't quite hold its shape, such as wax on a hot day. As the wax object prints, some melting areas might sag a bit more than expected. An open loop printer just keeps printing, resulting in a malformed object.

To correct problems like this, Daniel added an optical scanner to the 3D printer head. The optical scanner, in essence, watched exactly how things were printed. Daniel's closed loop printer, thanks to its scanner, could detect printing problems such as this sagging effect.

Just detection, however, was only the first step. By creating software that could read the output from the optical scanner, the 3D printer was responsive to changes in its environment. By keeping track of printing outputs, adapting the design to correct them, and adjusting its printing process in real time, you could argue that this particular 3D printer was "learning".

Closed loop printing contributes more than just monitoring the shape of printed output. A closed loop 3D printer, if programmed correctly and given the appropriate hardware to convey environmental feedback, could monitor the strength of material it is printing and add material if necessary. An intelligent, closed loop 3D printer could monitor the conductivity or elasticity of the material while it's printed. In fact, any material or structural property in the final 3D printed object that can be measured and modified in real time could be fabricated using an adaptive, closed loop 3D printer following instructions from a dynamical blueprint.

## Changing the shape of design tools

First we shape our tools, and then the tools shape us. Design for 3D printing reflects the influence of years of physical constraints, similar to the way that recorded music is packaged and sold. Do you remember vinyl record albums? An album was a collection of songs, each about 3 or 4 minutes in length. Since a vinyl album could hold only so much material, each album was made up of about ten songs, half on one side, half on the other.

There's nothing sacred about an album when you think about it. There's no deep aesthetic theory that dictates that periodically, an artist must simultaneously release a batch of several songs of mostly equal length and store them all on a single physical storage unit that's given a name and decorative cover art. Mozart certainly didn't comply with this standard. Yet our conception of recorded music is still shaped by vinyl discs that have been obsolete for decades now.

Music fans of a certain age think in albums. Designers think in terms of constraints imposed by once-limited computing power and the physical limitations of manufacturing machines.

Our capacity to imagine and 3D print physical objects will expand in tandem with design software's ability to express shape. Future design tools will be intuitive to use; some will be responsive to touch, to movement, to environmental conditions. Computers will become more fluent at depicting shapes and intelligent enough to solve design problems that we can't.

Humans will guide their computer's design strategies by feeding in data or pruning out unwanted design solutions and breeding more desirable ones. As Sivam Krish wrote on his blog, *Generative Design*, "the answer lies not in eliminating the human designer, but in assisting the designer by managing the constraints and the requirements that evolve throughout the design process."[2]

In bioprinting, for example, there's no design software for vascular systems. Someday doctors will apply generative design software to this problem. Rather than designing a complex vascular system for a kidney by directly describing its intricate branching shape, a doctor would specify a few basic rules. Then, to "design" the kidney's vascular network, the doctor would launch the generative design software to "grow" a filigreed network of new veins. Nature has produced an astonishing array of complex "designs" for plants and animals, even non-living objects like crystals and patterns in the sand. 3D printers are beginning to allow us to make complex structures, but the designs are still out of reach.

# 14 The next episode of 3D printing

Humans distinguished themselves from their evolutionary ancestors by making tools. Additive manufacturing technology may be the ultimate tool that will perhaps change human culture forever. With the unpredictable yet irresistible force of tsunamis, each new wave of improved tools of design and production have tipped off tidal waves of social change. This book barely scratches the surface of the emerging world of printed physical objects.

When I was a postdoc at Brandeis University in Boston, Massachusetts, my faculty advisor Jordan Pollack and I printed a complete working robot. It was a simple robot, but one whose entire body was designed and printed automatically. It was late 1999 when the first robot stepped out of the printer, and it took another 10 months until on a summer day in late August 2000 the story made it to the front page of *The New York Times*.

It was a bittersweet moment. First, the newspapers' editors made it clear to us that the only reason the story appeared on the front page was because the editors were desperate for news. That day—a sultry summer day in 2000—was one of the slowest news days on record. There was nothing—absolutely nothing—else to put on the front page. So "Robots Making Robots" it was.

But a more lingering emotion had nothing to do with sharing news of the breakthrough. The more lingering emotion was dissatisfaction, dissatisfaction that the robot wasn't entirely 3D printed. Its body and joints were printed, and that was an exciting step forward. But most of the other parts that make a robot a robot—its wires, batteries, sensors, actuators, and its "brain"—were manually assembled. There was still a long way to go before we could truly print a complete robot, fully assembled, batteries included. It would be a while before we can print a complete active system—no assembly required.

When this new generation of products emerge from the 3D printer, their debut may still not trump news of a brewing political scandal, but may hint at an even larger tsunami of change ahead.

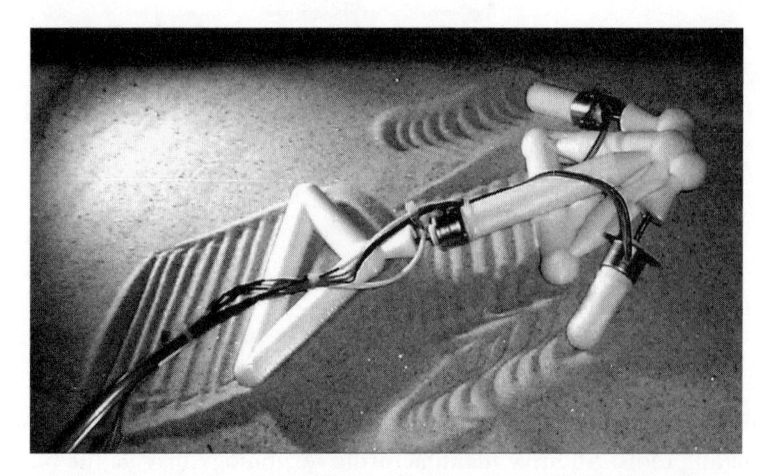

**The GOLEM project's printed evolved robot (2000). The white body was designed through evolutionary simulation and then fabricated using a 3D printer. Wires and motors were placed manually.**

Science fiction author Cory Doctorow said, "I'm of the opinion that science fiction writers suck at predicting the future. We mostly go around describing the present in futuristic clothes."[1] What are the futuristic clothes of 3D printing and design technologies? We'll soon have more material options, better printing resolution, faster fabrication speed, and lower machine costs. There will be new applications and unexpected new objects made. And after that?

Image courtesy of Jürgen Stampfl, Vienna University of Technology

**285-micron racecar: 3D printing at the microscale using stereolithography process**

## The three episodes of 3D printing

We started this book by looking at the evolution of additive manufacturing technology as a series of milestones that mark human progress in gaining control over physical matter. The first episode of this journey that's maturing today involves gaining unprecedented control over the *shape* of objects. 3D printers of today can already fabricate objects of almost any material—from nylon to glass, from chocolate to titanium, and from cement to live cells.

The ability to create arbitrary shapes is already having profound implications beyond engineering design. Mass manufacturing is becoming mass customization. In the future, as 3D printing technologies improve, everyone will gain the ability to design and make complex products. Barriers of resources and skill that are associated with traditional manufacturing will ease, democratizing innovation and unleashing the long tail of human creativity.

The second episode of the journey is just beginning: control over the composition of matter—going beyond shaping just external geometry to shaping the internal structure of new meta-materials with unprecedented fidelity.

Someday we will be able to make materials within materials. When 3D printers can blend raw materials in new ways we will witness the emergence of entirely new classes of materials. Materials manufacturing processes will shed traditional limitations which dictate that each part be made of a single material and then later assembled. With multi-material printing, multiple entangled components could be co-fabricated simultaneously, already pre-assembled. At a smaller scale, we will begin to embed and weave multiple materials into complex microstructures specified with micron-scale precision.

With such possibilities, you'll be able to print a custom tennis racket that cleverly amplifies your unique backhand or a replacement spinal disc implant tailored for your aching back (yet both will likely be unauthorized in professional sports). Although the possibilities are vast, few material scientists can predict properties of these new materials, and few designers can exploit the new design space. New design tools will be needed to augment human creativity.

The third episode of this journey, of which we are beginning to see early signs, is the control over behavior. In this episode we will go beyond controlling just the shape of matter as in episode one. We will move past controlling composition, as in episode two. In episode three, we will gain the ability to program materials to function in desired ways. We will move from printing passive parts and materials to printing integrated, active systems that can sense and react, compute and behave. We will move from controlling an object's mechanical functionality to controlling how it processes information and energy as well.

When this day comes, we'll be able to print (almost) anything—from a cell phone to a robot that strolls out of the printer. But like any sci-fi story, there's also a catch. That robot will not look at all like today's robots because it will not be limited by the constraints imposed by conventional manufacturing. Nor will it be designed directly by humans, because the new design space is too large for humans to fathom. The ability to manufacture active systems made up of both passive and active substructures with such freedom will open the door to a new space of designs and a new paradigm of engineering, one as powerful as biology.

## Cofabrication of multiple materials

At the core of the longer term future of 3D printing technologies is the ability to print with multiple materials simultaneously. In previous chapters we covered printers that print multiple components, each made of a different material. For example, a plastic extruder can print with red plastic one day and with white plastic the next. A metal printer can print titanium intertwined with stainless steel. Indeed, the ability to co-fabricate parts of different materials can remove challenges and limits of traditional assembly and enable us to make increasingly complex objects. But what makes multi-material printing truly exciting is the ability to co-print multiple materials simultaneously, patterning them together into complex, new meta-materials.

In the early days of paper printers, there were some dot matrix printers that had a quad-color ribbon that could print dots in red, green, blue, or black. However, you could choose only one color at a time. There were even pen plotters that could use eight pens with eight different colors but you had to load the plotter with the colors you wanted in advance, and each line could use only one pen.

The breakthrough came when printers could mix primary colors on-the-fly at increasingly precise and high resolutions. Like moving from a monochrome printer to a color printer, or from a black-and-white TV to a color TV, adding just three primary colors gives rise to millions of shades of color. The versatility of what you can make on a 3D printer will grow *exponentially* with the number of primary materials that can be printed and mixed simultaneously. This is because you can print not only the base materials, but also combinations of those primary materials that give rise to a combinatorial number of new permutations.

One of the companies pioneering that exploration of multimaterial printing is Objet—an Israeli company near Tel Aviv, that recently merged with Stratasys. I visited Objet headquarters to see what is yet to come. Located in a bustling science park next to orange groves, Objet is changing the way people think about materials. Objet's CTO Eduardo Napadensky and lead Material Scientist Daniel Dikovsky show me through the reception hall, laden with multi-material prints of anatomical models, industrial prototypes, and toys.

Daniel and Eduardo explained that multimaterial printing is not just about mixing materials. It's about creating new kinds of materials altogether.

When material scientists obsess over new materials, they are usually interested in new material properties—characteristics such as weight, strength, and flexibility. Often engineers are interested in combinations of properties, such as a material that is both light and strong, or both flexible and optically clear. Some material properties are intuitive, such as density and flexibility, and others are less intuitive, such as how many push-pull cycles a material can go through before it breaks or how long can it be stretched before it snaps. For example, when engineers design an aircraft wing, the push-pull cycles exerted on the wing structure by turbulence, or the stress on the fuselage associated with pressurization and depressurization cycles, need to be accounted for, a property known as fatigue strength.

Material properties can become complicated to understand and predict, and "materials by design" remain the Holy Grail of material science. 3D printing can greatly expand the materials that can be made. The problem is, however, that we don't know where to look and what to expect.

When first starting to print with multiple materials, one's intuitive impression is that properties of the combined mixed materials will lie somewhere between the properties of the base, primary materials. It stands to reason that if you mix a hard material with a soft material, in equal parts, you will get a semi-soft, semi-hard material. It turns out that this is only partially correct. The outcome of the resulting material depends on exactly *how* you mix the materials.

For example, if you print hard and soft materials in a checkerboard pattern, the result will be a new material that has certain strength. But if you print the same two materials, still in equal parts but this time in a random pattern, you get a much stronger material. The pattern, the way different materials are blended together by the printer, matters.

If you squint and look down a checkerboard pattern, for example in checkerboard floor tiling, you can see alternating black and white diagonals. Similarly, when two materials are printed in a checkerboard pattern, the long diagonal chains of soft material become a "weak link," which makes the composite material weak. But when you print the two materials in a random pattern, there is no such perfectly aligned weak link, so the overall material is much stiffer.

Scientists have long known that the nanoscale arrangement of atoms in a material matters to its macroscale properties. Practitioners figured out centuries earlier that random patterns are stronger than regular patterns: Blacksmiths quenched hot metal swords in cold water so that the metal cooled quickly and formed small crystals with lots of random boundary patterns, rather than annealing slowly into soft, smooth, malleable iron. But for the first time, we can *control* these patterns directly, explicitly. Not yet at the atomic nanoscale level, but at the microscale, gradually inching our way down.

Material properties can get even stranger. If you print multiple materials in certain patterns, their material properties can go outside the range of the base materials. You can mix weak and strong materials in a certain pattern and get a new composite material that is even tougher than either one of the original materials. It's a bit like getting steel-like material by mixing wood and plastic, both weaker than steel.

One of the things that makes a material weak is that small imperfections can turn into tiny cracks that then grow and propagate through the structure until it breaks. When we can strategically embed soft material inside a brittle material, those soft material patches can cushion the cracks and stop their propagation, delaying their catastrophic effect on the hard material, making it even tougher. Clamshells have interesting properties like this, but until recently, making such materials was the purview of Mother Nature. We certainly couldn't injection-mold materials like that. But with multimaterial printing, we might be able to fabricate them at will.

Here's another example. This one pertains to the elastic properties of materials and how they deflect and stretch under load. If you have ever stretched a rubber band, you will have noticed that as it gets longer, it also gets thinner. Most materials do that; it's called the Poisson effect, after the French mathematician and physicist Siméon Denis Poisson who first characterized it systematically. Yet it is possible to print hard and soft material in such a pattern that causes the material to *expand* laterally when it is pulled longitudinally. This bizarre, unnatural material property is a material with a negative Poisson ratio, also known as an *auxetic* material.

Auxetic materials are not found in nature and are difficult to manufacture using conventional manufacturing techniques. But with a high-resolution multimaterial 3D printer, you can make auxetic materials on demand and embed them within other structures to make strange and beautiful machines. For example, cars made with materials designed using auxetic materials to absorb energy upon impact so that the passengers are safe. The front bumper would absorb the impact and send the energy in different directions using patterns of auxetic and conventional materials.

Other patterns could have even more unusual and useful behaviors. It is possible, to print hard and soft materials in laminated patterns that make a material be flexible in one direction but stiff in a different direction. This property may not be so exciting by itself, until you realize that you can print objects with tailored elastic properties. For example a custom brace or implant could assist a patient after a knee injury by allowing that patient to bend their knee freely in one direction while supporting the knee in another direction. A custom glove could enhance a rock climber's ability to hang on protrusions.

Image courtesy of Daniel Dikovsky, Objet Inc.

**Multimaterial with a self-healing micro structure. The ball-and-socket releases at critical stress but can join again and recover the original pattern.**

During my visit to Objet, Eduardo and Daniel reached into a drawer and pulled out a few pieces of bizarre new materials. They showed me one piece they said was a printed self-healing material. They explained that this material can sustain stress up to a maximum point. Then, when the stress exceeds that limit, the material will give way but then can "heal" completely after the stress is removed.

Such a material is made by depositing raw material into tiny interlocking "balls and sockets." A material made of millions of interlocking components would remain flexible until the balls pop out of the sockets. If the material were compressed, the balls would snap back into place and the material would regain its original behavior.

When I examined Objet's novel material closely, it looked just like an ordinary gray plastic. I imagined how useful this ordinary, yet extraordinary printed material could be. If you've ever had your car bumper break off because you backed into a bush at the end of the driveway (speaking from experience), imagine how nice it would be if you could just push the cracked bumper back into place, and it would perfectly heal by itself.

Dynamic materials could change from stiff to soft depending on how much stress they are subjected to. Just like ground coffee will be hard as a brick when it is vacuum-packed, yet flow like a fluid when the vacuum is released, so-called jamming materials could be printed to change their stiffness in response to their environment.

The effect of patterning on material properties is far from intuitive. Material scientists and engineers spend careers trying to predict properties of even relatively simple composite materials like carbon fiber laminates. The advent of high-resolution multimaterial printers will open up such a vast new design space that it is difficult to anticipate the properties of the materials that will be possible, let alone exploit them for design. As the capabilities of 3D printers to deposit multimaterials expand, people will discover new materials sometimes by accident and sometimes by careful research. Just as new CAD tools are needed to help designers explore new shapes using new languages and design concepts, new design tools are needed to explore the new range of materials.

## Moving from printing passive parts to active systems

So far this chapter has discussed printing passive materials—hard or soft, elastic or stiff. Passive materials respond to their environments in a predictable mechanical way. In the future we will print active materials that act and react, sense, compute, and respond to their environment. The quest for printing active materials has had lots of fits and starts, as a result, the vast majority of whiz-bang 3D prints today are still of the passive sort—small and large, simple and complicated—but always passive parts.

The lowest hanging fruit in the area of active material printing is printing conductive materials—materials that conduct electricity. We already know how to print metals, and metals are good conductors, so what's the problem? The challenge is to print electrically conductive materials embedded inside a nonconductive material, such as copper wire wrapped in plastic insulation. If you can print conductors within nonconductors, you could print, say, robot parts with pre-assembled wiring, cell phone cases with complex custom antennas, prosthetic devices with built-in sensors, and whole new kinds of consumer electronic devices.

Printing conductive wires presents a double challenge—one that reaches beyond merely multimaterial printing. The challenge is to make sure the two materials are mutually compatible. If you try to print metals and plastics at the same time, the temperature needed to melt the metal will burn the plastic, making the two materials incompatible.

It is possible to find special kinds of conductive metals that have a low melting temperature compatible with plastic, but those materials are rare and difficult to use. Alternatively, it is possible to find nonmetallic conductors, like electrically conducting plastics, but those are not yet quite as conductive as metals. And so the search continues.

At Objet, Eduardo and Daniel were confident that printing electrically conductive materials embedded within nonconductive structural materials is within reach. The challenge is not technical, they insisted; it's a matter of business priority. Industry is now craving 3D printing materials that are stronger and more durable. Conductive materials are a bit beyond the short-term commercial horizon. They're not yet a priority.

There is a vicious cycle here: Industry wants stronger materials because it is seeking to emulate the existing materials and capabilities of traditional manufacturing technologies. The ability to fabricate embedded 3D conductive wiring does not currently exist in any form in traditional manufacturing. Except for a few avant-garde robot designers, a market does not yet exist for 3D printed wiring.

Academia's mandate, however, is to look beyond the short-term horizon. If you peer into the longer-term future, conductive materials will be just the tip of the iceberg of active materials. Printing batteries, motors and actuators, transistors, and sensors are just few of the possibilities being explored by researchers today.

When we look at printing active materials, we're actually talking about active systems. It is rare that an active material is useful by itself; it usually requires multiple active materials together, to make something useful. The challenge is moving from printing passive single-material parts to printing active, multimaterial integrated systems.

**Printed battery**

A good example of an integrated system is a 3D printed battery. If you open a battery chemistry textbook, you'll find dozens of recipes to make batteries—standard alkaline batteries, rechargeable lithium-ion batteries, zinc-air batteries, and lots more. All batteries have the same basic structure: an anode and a cathode material, with a "separation layer" in between, like a cheese sandwich with white bread on top and whole wheat bread underneath. The large, thin sandwich is then rolled up and encased in a tube and connected to two wires: one to the anode and one to the cathode.

What makes the battery work is that ions (charged atoms) present in the anode badly want to move into the cathode. When they move, they create a bit of electric current. Different combinations of anode and cathode materials (bread), separation layers (cheese), and battery geometry make for a vast range of battery types and performance characteristics.

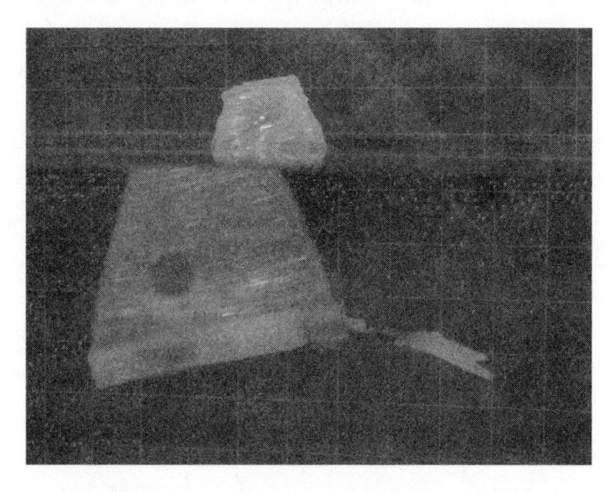

**Evan Malone's 3D printed robot fish, battery and actuator included (2010). It didn't quite swim out of the printer.**

Evan Malone joined our lab at Cornell shortly after the first robot appeared in the *The New York Times*. His goal was to print a robot that would walk out of the printer. Long before addressing any technical challenges, Evan had to address a different kind of challenge: none of the commercial printers available at the time would allow people outside the company to explore new materials. 3D printing companies closely guarded their patented materials and any attempt to use unauthorized materials would result in loss of warranty. Moreover, no reasonable researcher would stick active materials into a $100,000 machine only to see it jam up and stop working.

The quest to explore new materials unconstrained by existing commercial printer limitations (technical constraints and contractual restrictions) inspired the Fab@Home project (and to some extent also the RepRap project) and a series of open "hackable" 3D printer platforms that followed. Simple, low-cost DIY printers are more open to innovation and less of a tragedy to their user if they stop working.

Evan loaded a 3D printer with the relevant materials and quickly ran into the first snag: the anode material, zinc powder, turned into a paste and did not want to flow through a syringe tip. The harder he pushed, the more it resisted.

When that problem was overcome using various soaps and gels, Evan faced his next challenge. In most batteries, the recipe calls for a separation layer made of paper; even the commercial batteries you have at home likely have a paper layer separating the anode and cathode materials (that's the cheese between the two slices of bread). The separation layer can't be just any material—it has to be semi-permeable, allowing ions to flow through but not electrons. Paper is perfect, yet ironically, we could print almost anything, but not paper. After a few months of experimentation, Evan discovered a recipe for a printable separation layer made of a certain kind of gel.

Armed with a new recipe and five different materials, Evan printed a variety of batteries. Although their energy capacity was approximately one-half of optimized industrial batteries of equivalent size, their geometry was entirely custom. He could now print batteries in any shape he wanted, for example, in the shape of a leg for use as part of a robot.

Printing actuators—active material systems that can move—is even more challenging. We've printed electro-active polymer actuators, wax actuators, even electromagnetic actuators. The ultimate challenge, however, is putting it all together. An actuator alone is not quite enough, nor is an isolated battery. The ultimate milestone of active multi-material printing would be the creation of a complete robot that would walk out of the printer, batteries included.

## The final episode—from analog to digital

After the last episode, comes a an entirely new season—the biggest and most ambitious, the transition from an analog to a digital world. Before going further, we have to clarify something. The word "digital" has been overused. It can mean different things in different contexts:

1. Digital, meaning purely virtual, unembodied information—as in "physical versus digital," for example a digital newspaper
2. Digital, meaning electronic, programmable, as in "mechanical versus digital," such as a digital thermostat
3. Digital, meaning made of discrete, discontinuous units, as in "analog versus digital," such as a digital clock

Confusion arises because digital computers enable all three—they at once represent information virtually, electronically, and using discrete bits of ones and zeros. But it's possible to have "digital" objects that exist in physical form, yet maintain their digital nature in the other two senses of the word—they are programmable, and they are composed of lots of tiny, discrete bits.

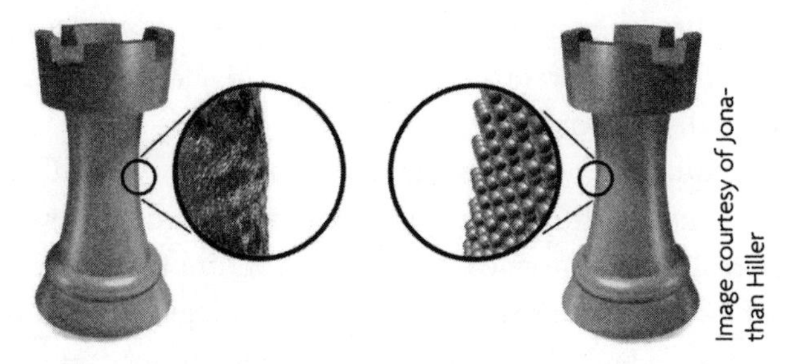


**Analog versus digital material. Digital materials are made of discrete physical bits called voxels.**

Most physical objects have an analog behavior. Analog systems are continuous, meaning that they transition smoothly—like the minute hand of a clock that moves smoothly through the infinite intermediate positions around the hour. A digital clock, however, does not move continuously. It has exactly 60 intermediate states, and it lingers on one state and then switches to the next instantaneously. Computer files are digital, in that they are composed of ones and zeroes. There's nothing in between.

In contrast, most current manufacturing techniques could be considered analog since produced materials are practically continuous. But that doesn't need to be the case.

I first met Neil Gershenfeld at his MIT Center for Bits and Atoms. Frankly, I was envious of the great name he chose for his center. I couldn't have come up with something that captures the essence of my own work any better. It was the summer of 2005 and we had just finished our first round of battery printing. The printed battery was "digital" only in the sense that it was electronically active, but it was physical and very analog in both other senses: It was composed of continuous streams of raw material.

I described our circuit printing effort at length. Neil's response was impatient. "Why not just deposit a chip with the whole prefabricated circuit in it?" he asked, as he reached into a drawer and pulled out a tiny transistor-chip, not larger than grain of rice. What if instead of droplets of ink, you deposited droplets of . . . circuits?

At first, I thought Gershenfeld missed the whole point. Dropping in prefabricated circuit components is a cheat—it defeats the whole point of printing circuits in the first place. But the more I thought about it, the more it made sense. Biological life is composed of 22 building blocks—amino acids—that arrange themselves in different permutations to give rise to a myriad of proteins and eventually life forms.

Biologists are quick to point out that there's much more to life than amino acids. Of course, living things need energy to put amino acids together and take them apart. But to a large extent, life's structure is composed of amino acid building blocks. This makes it possible for biological life forms to repair themselves. Animals and plants can consume each other and reuse the biomaterial because we are all made of the same relatively small set of just 22 building blocks.

In the same way a pixel is a building block of an image, a bit is a unit of information, and an amino acid is a building block of biological matter, a voxel is a volumetric pixel (hence its name). The elementary units of physical matter are atoms. The elementary units of printed matter would be larger, a couple hundred microns, the size of a grain of sand.

Image courtesy of Jonathan Hiller

**A rapid assembler, like a 3D printer, builds up objects layer by layer. But a rapid assembler does this by assembling very large numbers of very small building blocks.**

Like a few colors on an artist's palette, a few voxel types can take you far. If fewer than two dozen element types give rise to all biological life, a few basic voxel types can also open a large range of possibilities. To begin, let's combine rigid voxels and soft voxels. Using just those two types, of voxels, it's possible to make hard and soft materials. Add conductive voxels, to make wiring. Add resistor, capacitor, inductor and transistor voxels, to make electric circuits. Add actuator and sensor voxels and you have robots.

Voxels do not yet exist beyond the lab, and printers that can handle voxels do not yet work on a practical scale. But the idea that everyday objects will be made of billions of tiny voxels of a relatively small repertoire is mind boggling. Just as amino acids are the low-level common denominator that enables nature to recycle materials perfectly, if all products would be made of a few dozen basic voxel types, products could be "printed," then decomposed, and reprinted into other products.

To make this vision happen, we need to make tiny voxels, and find a way to rapidly assemble those voxels. A quick calculation shows that to make a small, shoebox size object from sand-grain sized voxels, you need approximately one billion voxels. Assembling one billion voxels can take a lot of time; even if a robot could perfectly assemble voxels at a rate of one per second, it would take almost 30 years. The solution is to assemble lots of voxels in parallel, a complete layer of voxels at a time, simultaneously.

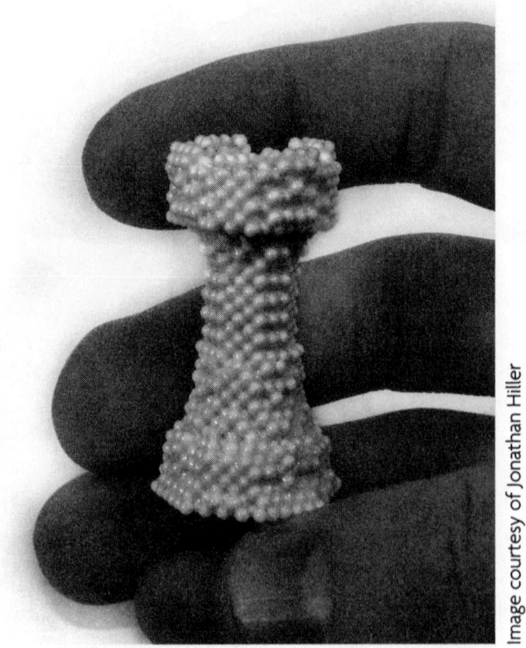

Image courtesy of Jonathan Hiller

**The first 10,000-voxel object assembled by a rapid assembler. Still coarse, like early computer graphics. Maybe one day we will have a GigaVoxel printer?**

Future assemblers may be able to pick up massive numbers of voxels in parallel (meaning several at one time) and place them into layers simultaneously, opening the door to what one day will be "digital materials." To distinguish these printers from their analog ancestors, we called these members of the new generation of machines "rapid assemblers." The next revolution after 3D printing will be the transition from analog to digital materials.

## Hybrid analog-digital printing

Imagine a future where human-made artifacts are composed of billions of tiny building blocks. These building blocks each have the same size and shape, and the same electric and mechanical connectors. Like tiny LEGO bricks, no larger than 100 microns in size—the size of a pixel on your screen—they interlock. Each of the two dozen or so building block types is made separately in bulk, and shipped as powder in a cartridge for an assembler.

Image courtesy of Robert MacCurdy

**Concept demonstration of hybrid digital-analog print. Body printed in transparent analog (smooth) material. Inside is a visible digital lattice structure of mock-voxels.**

The day when products will be made entirely of digital voxels may be far off, but meanwhile I expect that some combination of analog and digital materials will emerge. Hybrid 3D printing will combine continuous analog printing for some passive materials, and digital voxel printing for other materials that are more difficult to fabricate using continuous processes.

## Machines making machines

Technological singularity, a concept popularized by writer Ray Kurzweil, is a hypothetical future in which machines possess capabilities that enable them to accelerate their own development exponentially. One of the more widely recognized aspects of the idea of singularity is an "intelligence explosion" where intelligent machines design successive generations of increasingly powerful, even more intelligent machines.

As sophisticated as modern-day manufacturing machines have become, they do not have the capability to design and produce more physical instances of themselves. Today's manufacturing machines can't redesign themselves in response to some sort of challenge in their physical environment. 3D printing technologies will challenge our notions of what's normal, natural, or original by supplying a missing link in the speculation of singularity.

3D printers will someday fabricate active, digital matter that has the capability to reconfigure itself into intelligent machines that, in turn, will redesign and fabricate improved versions of themselves.

The idea of machines making machines is a recurring theme in both science fiction stories and serious academic studies. There are probably two reasons for this fascination: One is a practical view of scalability. Creating a machine that can create more machines leverages technology to its maximum capacity: With no humans in the loop, production is limited only by availability of material, power, and time.

The second reason for fascination with this topic may be rooted in a deeper psychological need—one that some might call hubris—our need to create. The natural distinguishes itself from the artificial in that natural creatures can make more creatures but machines cannot. At its core, self-reproduction is the ultimate hallmark of biology. If you can create a machine that can make other machines, you will have attained a new level of creation.

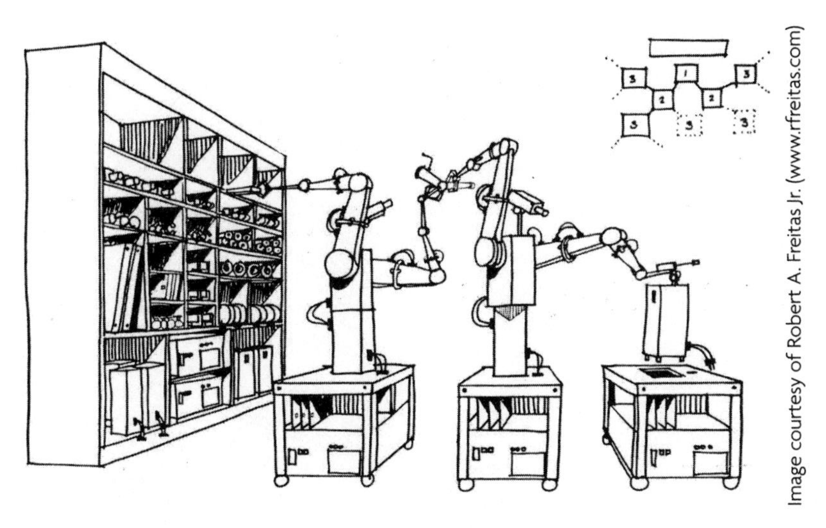

Image courtesy of Robert A. Freitas Jr. (www.rfreitas.com)

**Image from NASA-commissioned study, "A Self-Replicating, Growing Lunar Factory," Robert A. Freitas Jr. and William P. Gilbreath, AIAA, 1981**

In the future (or a long time ago in a galaxy far, far away) machines will make machines. 3D printers are the first wave of a new generation of machines that will design, make, repair, and recycle other machines. Machines will even adapt and improve other machines—and themselves.

I once took my son to watch one of the episodes of *Star Wars*. During several of the scenes, various spacecraft and robots equipped with lasers and other means of destruction were blowing up vast amounts of property—structures, vehicles, and other robots. Finally, after watching the destruction continue on and on, I mustered a grown-up demeanor and whispered loudly, "What a shame. So much effort spent by thousands of people flushed down the drain in a few seconds. It will take months if not years to rebuild all this."

My son had no clue what I was talking about.

"People didn't build this," he said. "Robots did."

# References

## Chapter 2

[1] Neal Gershenfeld, *When Things Start to Think* (New York, NY: Henry Holt & Company, 1999).
[2] Nicholas Negroponte, *Being Digital* (New York, NY: Random House, 1995).

## Chapter 3

[1] Quote from a press conference covered by VentureBeat in May 2012. http://venturebeat.com/2012/05/10/3d-systems-ceo-we-want-3d-printing-to-be-as-big-as-the-ipad/
[2] Quote from Terry's blog, July 2012. http://wohlersassociates.com/blog/2012/07/why-most-adults-will-never-use-a-3d-printer/

## Chapter 4

[1] Chris Anderson, *The Long Tail: Why the Future of Business is Selling Less of More* (New York, NY: Hyperion Press, 2008).
[2] Joseph Pine and James Gilmore, *The Experience Economy* (Boston, MA: Harvard Business School Press, 1999).
[3] Eric Reis, *The Lean Startup: How Today's Entrepreneurs Use Continuous Innovation to Create Radically Successful Businesses.* (New York, NY: Crown Publishing Group, 2011).
[4] http://en.wikipedia.org/wiki/Microcredit

## Chapter 5

[1] Harris L. Marcus, Joel W. Barlow, Joseph J. Beaman, and David L. Bourell, "From computer to component in 15 minutes: The integrated manufacture of three-dimensional objects." *JOM: Journal of the Minerals, Metals and Materials Society* 42, no. 4 (1990): 8–10.

[2] Paul Williams, "Three Dimensional Printing: A New Process to Fabricate Prototypes Directly from CAD Models." Master's thesis, MIT Mechanical Engineering, 1990.

[3] Christophe Chaput and J. B. Lafon, "3-D Printing Methods: 3-D Printing Based On Laser Stereolithography Opens Up New Application Fields for Advanced Ceramics." *Ceramic Industry* (2011): 15–16.

[4] Grant Marchelli, Renuka Prabhakar, Duane Storti, and Mark Ganter, "The Guide to Glass 3D Printing: Developments, Methods, Diagnostics and Results." *Rapid Prototyping Journal*, 17, no. 3 (2011): 187–194.

## Chapter 6

[1] John Walker, "The Autodesk File: Bits of History, Words of Experience," Fourth Edition (1994). Retrieved from `http://www.fourmilab.ch/autofile/`

[2] Richard P. Feynman, *Surely You're Joking, Mr. Feynman!* (Bantam, 1985).

## Chapter 7

[1] Russ Banham. "Printing a Medical Revolution." Connections. T. Rowe Price (May 2012).

[2] E. C. Armstrong (Ed.), *The Medieval French Roman d'Alexandre, Vol. 2* (Princeton, NJ: Princeton University Press, 1937).

[3] White House press release, June 4, 2012. `http://www.whitehouse.gov/the-press-office/2012/06/04/remarks-vice-president-joe-biden-cypress-bay-high-school-graduation-cere`

[4] Figures cited from Phil Reeves were gathered from Joris Peel's blog article, "3D printing in medicine: What is happening right now in patients," February 2011. `http://i.materialise.com/blog/entry/3d-printing-in-medicine-what-is-happening-right-now-in-patients`

[5] Figures cited from Phil Reeves were gathered from Joris Peel's blog article, "3D printing in medicine: What is happening right now in patients," February 2011. http://i.materialise.com/blog/entry/3d-printing-in-medicine-what-is-happening-right-now-in-patients

[6] Quote from Russ Banham. "Printing a Medical Revolution." *Connections*. T. Rowe Price (May 2012).

[7] Cyrille Norotte, Francois S. Marga, Laura E. Niklason, and Gabor Forgacs, "Scaffold-free vascular tissue engineering using bioprinting." *Biomaterials*, 30(30) (2009): 5910–5917.

[8] Emma Marris, "How to print out a blood vessel," *Nature* online. March 2008. http://www.nature.com/news/2008/080320/full/news.2008.675.html

[9] Miguel Castilho, Ines Pires, Barbara Gouveia, and Jorge Rodrigues, "Structural evaluation of scaffolds prototypes produced by three-dimensional printing." *Journal of Advanced Manufacturing Technology*, 56 (2011): 561-569.

[10] http://biomimicry.net/letter.html

[11] Quote from Russ Banham. "Printing a Medical Revolution." *Connections*. T. Rowe Price (May 2012).

[12] V. Mironov, T. Boland, T. Trusk, G. Forgacs, and R. R. Markwald, "Organ printing: computer-aided jet-based 3D tissue engineering." *Trends Biotechnol*, 21(4) (2003 Apr): 157-61.

[13] "Printing Living Tissues: 3-D Printed Vascular Networks Made of Sugar," *ScienceDaily*, July 1, 2012. http://www.sciencedaily.com/releases/2012/07/120701191617.htm

[14] Zhuo Xiong, Yongnian Yan, Renji Zhang, and Xiaohong Wang, "Organism manufacturing engineering based on rapid prototyping principles." *Rapid Prototyping Journal*, 11(3) (2005):160–166.

## Chapter 8

[1] Concept designs developed in collaboration with Amit Zoran. Prototypes developed with the assistance of Zachary Nelson, Josh Ramos, and Varun Perumal.

[2] http://web.media.mit.edu/~marcelo/cornucopia/

[3] Stephen Wolfram blog, "The Personal Analytics of My Life," March 8, 2012. http://blog.stephenwolfram.com/2012/03/the-personal-analytics-of-my-life/

[4] "4 Most Harmful Ingredients in Packaged Foods," *Reader's Digest* online. http://www.rd.com/health/diet-weight-loss/4-most-harmful-ingredients-in-packaged-foods/

[5] Rachel Laudan, "In Praise of Fast Food," *Utne Reader*, September-October 2010.

[6] Rachel Laudan, "In Praise of Fast Food," *Utne Reader*, September-October 2010.

[7] Marcel Dicke and Arnold Van Huis, "The Six-Legged Meat of the Future." *Wall Street Journal*, February 19, 2011. http://online.wsj.com/article/SB10001424052748703293204576106072340020728.html

## Chapter 9

[1] "High Schools: Are they doing their job?: Much criticism. Much anxiety. What's the truth?" *Changing Times* 10 (1956): 27–29.

[2] NSF Report, Horwitz, 1995

[3] Frank Coffield, David Moseley, Elaine Hall, and Kathryn Ecclestone, *Learning styles and pedagogy in post-16 learning: A systematic and critical review*. (Published by the Learning and Skills Research Centre, 2004).

[4] Frank Coffield, David Moseley, Elaine Hall, and Kathryn Ecclestone, *Learning styles and pedagogy in post-16 learning: A systematic and critical review*. (Published by the Learning and Skills Research Centre, 2004).

[5] Harold Pashler, Mark McDaniel, Doug Rohrer, and Robert Bjork. "Learning Styles: Concepts and Evidence." *Psychological Science in the Public Interest* 9(3) (2008).

[6] Glenn Bull, Gerald Knezek, and David Gibson, "A rationale for incorporating engineering education into the teacher education curriculum." *Contemporary Issues in Technology and Teacher Education*, 9(3) (2009): 222-225.

[7] Cheryl A. Cox and John R. Carpenter, "Improving attitudes toward teaching science and reducing science anxiety through increasing confidence in science ability in inservice elementary school teachers." *Journal of Elementary Science Education*, 1(2) (1989): 14-34.

## Chapter 10

[1] "MIT Looks at Printing Buildings." http://fabbaloo.com/blog/2011/4/13/mit-looks-at-printing-buildings.html
[2] Stephen Todd and William Latham, *Evolutionary Art and Computers.* (Academic Press, September 1992).

## Chapter 11

[1] China Labor Watch reports. http://www.chinalaborwatch.org/
[2] "ATKINS: Manufacturing a Low Carbon Footprint. Zero Emission Enterprise Feasibility Study." Project number N0012J, October 2007. Lead Partner: Loughborough University.
[3] "ATKINS: Manufacturing a Low Carbon Footprint. Zero Emission Enterprise Feasibility Study." Project number N0012J, October 2007. Lead Partner: Loughborough University.
[4] Whitney MacDonald, "Time for Titanium Processing." *CSIRO Process* magazine (June 2005): 1-2. http://www.csiro.au/files/files/p81m.pdf
[5] "ATKINS: Manufacturing a Low Carbon Footprint. Zero Emission Enterprise Feasibility Study." Project number N0012J, October 2007. Lead Partner: Loughborough University.
[6] http://en.wikipedia.org/wiki/Recycling
[7] Susan Freinkel, *Plastic: A Toxic Love Story* (New York, NY: Houghton Mifflin Harcourt Publishing, 2011).
[8] Stat from Global Environmental Polymers, Inc. http://www.degradablepolymers.com/plastic_pollution.html
[9] Capt. Charles Moore with Cassandra Phillips, *Plastic Ocean: How a Sea Captain's Chance Discovery Launched a Determined Quest to Save the Oceans* (Avery, 2011).
[10] Joris Peels, "3D printing vs Mass Production: Part IV More beautiful landfill." i.materialise blog (June 29, 2011). http://i.materialise.com/blog/entry/3d-printing-vs-mass-production-part-iv-more-beautiful-landfill
[11] American Chemistry Council, "2005 National Post-Consumer Plastics Bottle Recycling Report" (2005).

## Chapter 12

[1] United States Secret Service. "Know Your Money." http://www.secretservice.gov/money_technologies.shtml

[2] Sebastian Anthony, "The world's first 3D-printed gun." *ExtremeTech* (July 26, 2012). http://www.extremetech.com/extreme/133514-the-worlds-first-3d-printed-gun

[3] Mark D. Symes, Philip J. Kitson, Jun Yan, Craig J. Richmond, Geoffrey J. T. Cooper, Richard W. Bowman, Turlif Vilbrandt, and Leroy Cronin, "Integrated 3D-printed reactionware for chemical synthesis and analysis." *Nature Chemistry*, 4 (2012): 349-354. doi:10.1038/nchem.1313

[4] Nikki Olson, "3D Printing Laboratories: The Age of DIY Designer Drugs Begins." Institute for Ethics & Emerging Technologies (April 26, 2012). http://ieet.org/index.php/ieet/more/olson20120426

[5] Peter Hanna. "The Next Napster? Copyright questions as 3D printing comes of age." *Ars Technica* (April 6, 2011). http://arstechnica.com/tech-policy/2011/04/the-next-napster-copyright-questions-as-3d-printing-comes-of-age/2/

[6] Erin McCarthy, "SXSW: The Looming Legal Battles over 3D Printing." *Popular Mechanics* (March 14, 2012). http://www.popularmechanics.com/how-to/blog/sxw-the-looming-legal-battles-over-3d-printing-7333888

[7] http://en.wikipedia.org/wiki/Patent

[8] Simon Bradshaw, Adrian Bowyer, and Patrick Haufe, "The Intellectual Property Implications of Low-Cost 3D Printing." *SCRIPTed*, Volume 7, Issue 1 (2010). http://www.law.ed.ac.uk/ahrc/script-ed/vol7-1/bradshaw.asp

[9] Phillip Torrone, "MAKE's Exclusive Interview with Alicia Gibb – President of the Open Source Hardware Association." *Make* magazine (April 23, 2012). http://blog.makezine.com/2012/04/23/makes-exclusive-interview-with-alicia-gibb-president-of-the-open-source-hardware-association/

[10] Mark A. Lemley and Carl Shapiro, "Probabilistic Patents." *Journal of Economic Perspectives*, Volume 19, Number 2 (Spring 2005): pp. 75–98.

[11] http://www.crnano.org/dangers.htm

## Chapter 13

[1] Stephen Todd and William Latham, *Evolutionary Art and Computers*. (Academic Press, September 1992).

[2] `http://generativedesign.wordpress.com/2011/12/12/cracking-the-layout-problem/#more-1719`

## Chapter 14

[1] Katy Scott, "Where is my flying car?" *3rd Degree*, 4(10) (September 2007). `http://3degree.ecu.edu.au/articles/1378`

# Index

## B

## C

## D

## E



## F

## N



## O

## X–Y–Z